IEE HISTORY OF TECHNOLOGY SERIES 23

Series Editor: Dr B. Bowers

History of international broadcasting

Volume 2

Other volumes in this series:

History of international broadcasting

Volume 2

James Wood

Published by the Institution of Electrical Engineers
in association with the Science Museum

Published by: The Institution of Electrical Engineers, London,
United Kingdom

© 2000: The Institution of Electrical Engineers

The Institution of Electrical Engineers,
Michael Faraday House,
Six Hills Way, Stevenage,
Herts. SG1 2AY, United Kingdom

British Library Cataloguing in Publication Data

A CIP catalogue record for this book
is available from the British Library

ISBN 0 85296 920 1

Printed in England by Bookcraft (Bath) Ltd

Cover: The massive SW curtain array at VOA, Delano against an azure sky depicts
the upper reaches of the ionosphere. Inset pictures clockwise from left, Gateway
satellite earth station, VOA Delano CA; First ALLISS nearing completion
at RFI SW site Issoudun, France 1993; RF tuning mechanism of Telefunken S 4105
500 kW SW transmitter. All pictures reproduced by kind permission of: TCI
Sunnyvale, CA; Voice of America, Washington DC; Thomcast, Conflans Ste
Honorine, France andTelefunken Sendertechnik, Berlin.

'Too many people expect wonders from democracy, when the most wonderful thing is just having it.'

Walter Winchell

Contents

List of figures

Preface

As one who was born in the year that station KDKA Pittsburgh inaugurated the age of broadcasting in the US, my love of radio was fired at the age of ten when my home-built wireless receiver picked up its broadcasts. As an engineer I have been fortunate to have been involved with some of the major radio events of the twentieth century. Writing on the role of broadcasting and communications in these events, and the technology itself, has rewarded me with worldwide travel, friendships and intellectual conversation with many like-minded fellow veterans.

In my previous book *History of international broadcasting* I attempted to encompass the history of international radio broadcasting in a single volume. However, history is constantly being made and, momentous as the four decades of the cold war were – with shortwave broadcasting emerging as a major force – that history is now matched by what has since happened in the former Soviet Union and some countries in Eastern Europe. During the cold war no one could have predicted that in the not-too-distant future Western broadcasters such as Voice of America, Deutsche Welle, the BBC World Service and at least one religious broadcaster would be leasing airtime on Soviet-built transmitters and have access to some of the former Soviet Union's most powerful SW transmitter facilities in the world, which were once used for jamming Western broadcasts. Yet this is exactly what has happened and these Western broadcasters are now able to target China and other countries in Asia with increased audibility.

In my previous book I expressed the view that wars, whether hot or cold, act as a spur to advancements in technology and science. This is especially true of SW broadcasting. The collapse of the Soviet empire, in which Radio Free Europe/Radio Liberty played a key role, is one such example. Through a policy of innovative engineering, RFE/RL achieved the missions that it had been set. But RFE/RL engineers did much more; they elevated the art and science of SW broadcasting to unprecedented heights. As a Moscow listener observed, "There was never an audibility problem with Radio Liberty, it came in like a local radio station."

Today, the United States is the sole surviving superpower so it seems right that much of this volume is given over to the US – its aspirations and global strategies, its broadcasting arms, VOA, RFE/RL, the Board of International Broadcasting (BIB) and the newly structured International Broadcasting Bureau (IBB). Where my first volume was not at liberty to go into detail on the technical structure of RFE/RL SW stations, *Volume 2* is able to do so, and at the same time to pay tribute to the achievements of this unique broadcaster.

Other chapters in this book are devoted to a number of well-respected international broadcasters not included in the first volume. They include Radio Canada International, Radio Netherlands International and Swiss Radio International. Broadcasters from the Arab/Islamic world, stretching from Morocco, through North Africa, to the Gulf Region and the Middle East, are included, as are some broadcasters from Asia especially SE Asia and the Pacific rim.

Although SW continues to be the main gateway for the projection of foreign policy, there has been some resurgent interest in high and very high power MW, particularly in the Middle East and Europe. The medium waveband has propagation qualities which make it nearly perfect for medium distance, cross-border broadcasting.

This resurgence of interest has come about primarily through the efforts of the Harris Corporation whose Broadcast Division developed the world's first high efficiency all-solid state technology that makes it possible to construct transmitters with carrier powers of 1 to 2 megawatts. The numbers of units sold to broadcasters provides an example of how the right product can actually generate a market. Analysis of world sales for high power MW transmitters indicates a growing market from broadcasters who have come to appreciate the pay-back that results from savings in electrical power consumption from the 83–86 per cent efficiency offered by the newest generation of AM transmitters.

As for the SW market, the end of the cold war has caused a certain amount of confusion amongst some international broadcasters. Some have made cut-backs in programme hours, blaming this on a reduction in world listening audiences, whilst at least one broadcaster – the BBC World Service – claims an audience increase to over 130 million. The religious international broadcasters also take a very positive view of the value of SW. Two 500 kW transmitters can project the Christian Gospel to the Continent of Africa which do what in the nineteenth century would have taken several thousand missionaries. Christian religious broadcasters such as AWR, FEBC and TWR play powerful roles in the broadcast world of today, and set examples to lesser international SW broadcasters that makes reporting their operations appropriate in this volume.

There has been a constant upward trend in output powers of SW transmitters over the past two decades, but with 500 kW carrier power equal to 2000 kW peak envelope power, such powers are nudging closer to the limits of physics, and while it is possible to combine two 500 kW SW transmitters, I believe we are more likely to see greater emphasis on SW curtains with higher gain and increased directivity. Chapters on recent developments in transmitters and on curtain arrays with rotatable designs are included in this volume. In this context it may be noted that recent years have seen some company mergers, acquisitions and, in a few cases, collaborative ventures in an attempt to cope with a market where research and development costs are rising but volume sales of transmitters has slowed, creating a cycle of diminishing returns on capital investment.

SW as a delivery system has come under examination in recent years with regard to its long-term future in the light of emergent technologies such as digital radio broadcasting. The use of a satellite system has been proposed and Worldspace, a US company, is likely to become operational in 1999 with limited coverage. However, I believe this system is more of an adjunct to SW, rather than a replacement for it. As a delivery system, amongst other disadvantages it suffers from the 'goalkeeper' effect, in that the owner of the satellite(s) is a third party with powers of veto; in a potential world crisis it could refuse to carry a certain programme. In this respect, SW offers a seamless, uninterruptable flow of information from broadcaster to listener, with no censorship or other third party involvement.

SW is not standing still in the technology race. Some significant developments have taken place, and it is now possible to transmit digital audio sound in the AM wavebands. It will retain all the advantages of AM broadcasting but with the added benefit associated with transmitting a digital signal.

Acknowledgements

As any good journalist and historian knows, the problem with writing an historical account is not usually one of having insufficient material, but rather of deciding what has to be excluded for reasons of space. This *Volume 2* of *History of international broadcasting* is no exception. The writing of a technical and political history that has spanned one hundred years in a few hundred pages has inevitably meant that much interesting material has had to be omitted, not because that material is unimportant but because it did not fit into the theme of this volume. There are technologies in this book which warrant their own history, but I have had to restrict coverage to the part they play in international broadcasting in the HF spectrum.

I have chosen to include material on some important international broadcasters, both government and religious, who were not in the first volume, for example Radio Canada International, Radio Netherlands International, Adventist World Radio and the Far East Broadcasting Company. To those who are not included I offer my apologies in advance. In researching the book since 1992 I have met with co-operation, kindness and in some instances generosity from broadcasters and manufacturers alike when it came to arranging visits to high power transmitter sites of a sensitive nature, and to the production plants of the major manufacturers in America and Europe. Whenever requests were made for information or facilities I rarely encountered difficulties. And so, to those concerned, I extend my appreciation.

I would like to thank Daniel Bouchant of RFI, Detlef Braun of Deutsche Telekom, Lisa Breeze of Radio Australia, Joseph O'Connell of VOA, Richard Baker of GEC-Marconi, Jim R Bowman and Natalie J Hike of FEBC International, Jacques Bouliane of Radio Canada International, Walter Fankhauser of SRI, Kevin Klose, President of IBB, Gregoric Leupold of RTO Slovenia, Ulrich Kundig of SRI, Fiona Lowery of the BBC World Service, Dr Graham Mytton and Carol Forester of the BBC's Audience Research unit – an unrivalled source of data, Jonathon Marks of RNW (Radio Netherlands International), Anthea McNeil of FEBA UK,

André Roumejon of TDF France, Andrew Larquie of Radio France International, Anthea and Allen Steele of AWR Europe, Peter Senger and Dieter Weirach of Deutsche Welle, Dr W Schminke of Thomcast, W Tschol of Thomcast AG Switzerland, and Johann Strohmayer of Deutsche Telekom, Augsburg.

Also, Guy Noel Le Carvennec, Patrick Bureau, Jean Charles Daninos and others of Thomcast France, formerly Thomson-CSF Gennevilliers, are thanked for their friendship and for visits to Thomcast plants and various transmitter sites in France; equally Jurgen Reich of Thomcast, Mannheim, Germany and Darko Cvjetko, President of RIZ, Oliver Sablic and Marijan Kassal for visits to the RIZ factory, Zagreb and other sites, especially Stefica Mahalup for her hard work, friendship and hospitality.

I thank Charles Kalfon, Operations Director of Thomson Tubes TTE, William R House, Guy Clerc, Head of Thomson Tubes Thonon, for the visit to a power grid tube plant, and for improving my knowledge of tubes. Also on tubes, my thanks to George Badger, once of EIMAC and now President of Svetlana tubes, for his unrivalled experience of the American tube industry and his insights into Russian tube science. I am grateful to Jurgen Graaff, Managing Director of Telefunken Sendertechnik, Berlin, for his friendship and co-operation in providing technical information and also for the history of Nauen in Chapter 7.

On antennas I am indebted to TCI Sunnyvale CA, and especially Dr R W Wilensky, Dr Ahmad Ghaffarian and Shaun Metcalfe, Chairman and MD of Technology for Communications International (UK) and also for being accorded privileges to inspect the world's largest curtain array in conjunction with VOA at the Delano, CA SW facility. I am indebted to Matthew Richards and Dr Ghaffarian for their excellent company on that visit who, along with Stephen Kershner, Designer of SW curtain arrays for RFE, RL and VOA, provided material for Chapter 24. Prof. Dr Dietmar Rudolph, Deutsche Telekom Engineering College Berlin, provided some of the information in Chapter 26. Russell E Geiger, former Director of Engineering, Radio Free Europe (1951–1966) helped me with RFE early history and George Woodard of IBB (former Vice-President of Engineering at RFE/RL) supplied data on the RFE/RL broadcasting network and transmitter sites. My thanks are also due to Robert Kamosa of VOA for his regular updates on the VOA Modernisation Programme.

I am grateful to many at Harris Broadcast Division, Quincy, Illinois: Martha Rapp, Jack O'Dear, Robert Weirather, Hilmer Swanson and, especially, Tom Yingst (former Vice-President) for the visits to the Harris plant – for his unfailing friendship and his unique skills in infusing others with his enthusiasm for superpower broadcasting in the AM wavebands.

I have drawn on articles which I have published in *Radio World* in my capacity as a columnist on SW transmission and international broadcasting and would like to thank Alan Carter, Editor-in-Chief, and Steven B Dana, Publisher at IMAS Publishing.

I am indebted to Lawrence J Cervon, veteran of the US broadcast industry, former President of Broadcast Electronics Inc., for his help in reading the manuscript, his constructive suggestions and also for providing some original archival material on the history of the Gates Radio Company.

As with the first volume of *History of international broadcasting* this second book includes material from personal archives and from my experience as a consulting engineer in the subject itself. I would like to thank the publisher, the Institution of Electrical Engineers, for commissioning this second volume, in particular Dr Robin Mellors-Bourne, Director of Publishing, and Jonathan Simpson, then Commissioning Editor, who both also supplied editorial assistance. Thanks also to Sarah Daniels, the Book Production Editor for her hard work during the production process.

Abbreviations

ABB	Asea Brown Boveri
AFRS	Armed Forces Radio Service
AFRTS	Armed Forces Radio & Television Service
AM	amplitude modulation
AOR	Atlantic Ocean Relay satellite
AT & T	American Telephone and Telegraph
AWR	Adventist World Radio
BBC	The British Broadcasting Corporation (also Brown Boveri Company)
BBC-WS	BBC World Service
BIB	Board of International Broadcasting
BSE	Broadcast Systems Engineering
CBC	Canadian Broadcasting Corporation
CD	compact disc
CEC	Continental Electronics Corporation
CIA	Central Intelligence Agency
CIS	Commonwealth of Independent States
CNN	Cable News Network
CRI	China Radio International
CW	continuous wave
DAB	Digital audio broadcasting
DAM	dynamic amplitude modulation
DBP	Deutsche Bundespost
DPRK	Democratic People's Republic of Korea
DRM	Digital Radio Mondiale
DSB	dual sideband
DW	Deutsche Welle
EBU	European Broadcasting Union
ERTU	Egyptian Radio & Television Union
FCC	Federal Communications Commission
FCO	Foreign & Commonwealth Office

FEBA	Far East Broadcasting Association (a Division of FEBC)
FEBC	Far East Broadcasting Company
FM	frequency modulation
FSU	former Soviet Union
HEO	highly elliptical orbits
HF	high frequency (HF and SW are interchangeable)
HFCC	High Frequency Coordination Committee
HRTV or HRT	Hrvatska Radio & Television (Croatia)
IBAR	International Broadcast Audience Research Unit (of the BBC)
IBB	International Broadcasting Bureau
IOR	Indian Ocean Relay satellite
IRIB	Islamic Republic of Iran Broadcasting
ITU	International Telecommunications Union
KBS	Korean Broadcasting System
KRT	Korean Radio & Television
LSB	lower sideband
LW	long wave
MF	medium frequency
MW	medium wave (MW and MF interchangeable)
NAB	National Association of Broadcasters
NAO	National Audit Office
NHK	Nippon Hoso Kyokia (Japan)
NRB	National Religious Broadcasters
OWI	Office of War Information (US)
PDM	pulse duration modulation
POR	Pacific Ocean Relay satellite
PSB	public service broadcaster
PSM	pulse step modulation
PTT	Post, Telephone & Telegraph
RBI	Radio Berlin International
RCI	Radio Canada International
RF	radio frequency
RFA	Radio Free Asia
RFE	Radio Free Europe
RFE/RL	Radio Free Europe/Radio Liberty
RFI	Radio France International
RIZ	Radio Industries Zagreb
RL	Radio Liberty (once Radio Liberation)
RMI	Radio Moscow International
RNW	Radio Nederland Wereldomroep
RTL	Radio-Television Luxembourg
RTV-SLO	Radio & Television Slovenia
SBC	Swiss Broadcasting Corporation
SFR	Socialist Federation Republic (of Yugoslavia)

SIS	satellite interconnect system
SRI	Swiss Radio International
SSB	single sideband
SW	shortwave (the terms SW and HF are interchangeable)
TCI	Technology for Communications International
TDF	TéléDiffusion de France
TOA	take-off angles
TWR	TransWorld Radio
UHF	ultra high frequency
USB	upper sideband
USIA	US Information Agency
VHF	very high frequency
VOA	Voice of America
VOFC	Voice of Free China
WARC	World Administrative Radio Conference

Chapter 1
International broadcasting in the HF spectrum: past and present

Since the end of the cold war and the abandonment of communism by the former USSR, there seems to have developed a feeling by some in the broadcast transmission industry in the West that shortwave (SW) broadcasting has seen its best days, and that developments in satellites, and more specifically the prospect of satellite delivery of digital audio sound broadcasting, will eventually spell the death of SW broadcasting. To assess why SW broadcasting is attracting some less than optimistic comment, and to consider what the future holds, it is necessary to look at what the medium has achieved in its sixty-plus years of history.

To understand what international SW broadcasting in the HF spectrum is about, it is useful to appreciate the politically unstable Europe of the mid 1930s onwards. This was the time when the most powerful nations of Europe realised that SW broadcasting was destined to become the most powerful tool for the projection of propaganda internationally. As a medium it could reach out beyond a nation's territorial borders, its contents could not be censored, nor could free flow be interrupted. Moreover, to hear such broadcasts required no great investment on the part of the listener other than a simple SW receiver.

SW broadcasting has proved itself as a potent tool in times of political crisis. During the build up to World War II, and throughout that war, propaganda played a dominant role, but this was merely a prelude to the much bigger role it would play in the cold war. Many people believed that the cold war could end in nuclear war. That it did not is in some part due to the role played by Radio Free Europe/Radio Liberty (RFE/RL), supported by the big three international broadcasters, the BBC World Service (BBC-WS); Voice of America (VOA) and Deutsche Welle (DW). In the words of the chairman of the US President's Task Force on international broadcasting, 'They sent out words, not bullets; ideas, not bombs, and they broke down a wall and helped break up an evil empire'. 'US taxpayers' money spent funding the operations of VOA and RFE/RL was one of the best investments that America had made.' [1]

1

More than 150 countries in the world now possess an international broadcasting capability in the SW bands. About two dozen or so are significant 'players', in addition to the 'big three' mentioned above. By accepted definition all are propaganda broadcasters because their role is to broadcast information in the interest of the originating state and to reflect government policy in their programmes. Specifically these international broadcasters are a tool for projecting a nation's foreign policies.

For all the expenditure by countries around the world on the medium, SW broadcasting does not attract a high profile, often due to deliberate policies on the part of governments: to promote SW would in turn make citizens of a country vulnerable to SW broadcasts from another country. In short we have the situation where countries want to export international broadcasting, but do not want to receive it. So, the popularising of SW listening is left to the independent and private amateur radio clubs and societies, and figures for SW listening in the countries of Western Europe, for example, are about 25 in every 1000 population. Nevertheless, there are regions of the world where SW listening is more popular, and local conditions or events can affect the profile of SW. For example, the 1991 Gulf War saw a dramatic increase in SW popularity and in sales of portable SW receivers.

Following the collapse of communism in the former USSR there was a decline in the number of tenders issued for new SW stations, a situation that was to be expected, perhaps, as countries took time to re-adjust to a major upheaval in world politics. The backdrop to this is that for 40 years – from the onset of the cold war to its abrupt end in 1990 – sales of high power SW transmitters boomed. SW transmitters became more powerful, output power climbed from 50 to 100, then 250, and finally to 500 kW – the so-called super transmitters of the 1980s. Throughout this period all the major broadcasters were enlarging existing transmitter complexes and embarking upon even bigger SW projects. In 1986 VOA issued a tender and RFP (request for prices) for a quantity of 100 'super transmitters'. (This quantity was later downgraded to 55.)

When the end of the cold war came, VOA, the BBC-WS and DW discovered a way to significantly increase listening audience figures without having to buy additional SW transmitters. Citizens of the former Eastern bloc countries and the peoples of the former republics of the USSR were eager to hear broadcasts from the West and to meet this demand Western broadcasters were quick to make agreements with local and state authorities which enabled their SW broadcasts to be re-transmitted over local and regional AM and FM radio stations.

To the governments of Britain, Germany and the USA this was an unexpected bonus which enabled them to broadcast to 350 million more people. The BBC alone estimated it gave them an additional 30 million more listeners.

The second bonus came to Western international broadcasters when they discovered that hard currency would enable them to lease what are now known to be the most powerful SW transmitters of their type anywhere in the world, with output powers of 1000 kW. These transmitters, installed deep in what were previously the most secret or restricted areas of far east Russia and in other former Soviet republics, had played a vital role during the cold war by jamming incoming SW signals, but now, either surplus to requirements or because the authorities could no longer afford the huge electricity bills for the upkeep of these sites, they became available to Western broadcasters.

It was an arrangement that suited all parties. The former Soviet republics were financially strapped for cash and hungry for dollars and Deutsche marks. To Western governments it was a heaven-sent opportunity to gain access to these extremely powerful SW transmission sites. Powerful 1000 kW SW transmitters at Novosibirsk, Dushanbe, Tajikistan and Kamo, along with several 500 kW SW transmitters at Irkutsk, Taskent, Chita, Novosibirsk, Krasnador and Petropavlovsk-Kamchatskiy, now began to carry Western broadcasts, not to the former Soviet peoples but to China, Laos, Cambodia and elsewhere. These were previously difficult regions to reach with good audibility, but the new short route and higher power changed all that.

What this story reveals is that SW broadcasting, far from dying apace, is being redeployed, while remaining as popular as it ever was in many regions of the world. Why else would VOA, the BBC-WS and DW be spending tax-payers' money on leasing now surplus high power and super power SW transmitters in east Asia and elsewhere? The downside to this falls to transmitter manufacturers, as for every Soviet-built transmitter that Western governments lease, and I estimate that there are about a hundred or so 'leased' transmitters in the former Soviet republics at the time of writing, the BBC-WS, DW and VOA need less transmitters from Western manufacturers.

The market for new transmitters in some regions of the world is more active, as some regions are better reached from other parts of the world. The BBC-WS, for instance, has improved its signal audibility in SE Asia, with a new SW relay station in Thailand. It is equipped with four of Thomcast's latest generation, one-tube 250 kW transmitters of type TSW 2250. The first of these transmitters was completed at Conflans Sainte Honorine, France at the end of 1995. Also high on the BBC's capital projects programme is the re-equipping of its East Asia relay station on Masirah Island in the Indian Ocean, presently equipped with a 1500 kW MW transmitter of 1973 design and four 100 kW SW transmitters supplied by Harris in 1976.

Radio France International (RFI), another of the big international broadcasters, led the field in 1992 when it announced the biggest SW transmitter expansion programme in Europe, the ALLISS project. When this programme was finished in 1997 France gained a SW broadcasting capacity second only in size to VOA and thought by many experts to be superior in its technology. See Chapter 8 for more information on ALLIS.

Another European international broadcaster, Deutsche Welle, completed a new SW transmitting station in 1996 at Nauen, 30 km outside Berlin in the former German Democratic Republic. This is fitted with four of Telefunken's 500 kW SW S-4105 transmitters together with four ALLISS rotable antennas, built at Thomcast's plant in Mannheim, Germany. A third European international broadcaster has also acquired new SW transmitters. This is Spain's foreign service broadcaster, Radio Exterior de Espana. Under a deal with RFE/RL, Spain received two 250 kW SW TSW 2250 transmitters.

Programme preparation costs

If the 1990s transmitter expansion projects of Europe's most powerful international broadcasters indicate that SW broadcasting will continue to play a key role in foreign affairs, then an examination of programme operating costs will reinforce that conclusion. The fundamental difference between foreign service broadcasters and internal broadcasting is that the former is an information broadcaster whereas national broadcasters are primarily concerned with entertainment. The cost of preparation and translation into the language or dialect of the target zone is often extremely high.

If we take the BBC World Service as one example, it broadcasts in English and a further 41 different languages. A recent UK National Audit Office study of programme language transmission costs has shown these as ranging from about £100 per hour to as much as £2123 per hour. The reason for the steep costs is the need for the highest possible accuracy in the projection of news and analytical material. Without accurate translation, broadcasts can become jumbled and meaningless or at least distorted, sometimes a serious passage can result in distortion of fact. BBC figures show that the highest expenditure and costs are associated with translations into Ukrainian, Albanian, Swahili, African French, Hausa, Somali, Slovak and Urdu.

It so happens that the highest translation costs are associated with broadcasts that are specifically targeted to some of the poorest countries in the world, where GDP and income per capital is low. Albania, for instance, is the poorest country in Europe, yet the BBC-WS dropped its service to France so as to make way for a language programme to the peoples of Albania. Major international political broadcasters such as the BBC-WS do not as a general rule target those countries with which their country enjoys a stable relationship, or those which are politically stable (and usually right wing) regimes. For example, Britain does not broadcast on SW to Japan these days, but did when it was at war with Japan in 1941.

Facts such as these, when coupled with the high investment which is still in progress with the construction of foreign service international broadcasting projects around the world by major international broadcasters, provide sufficient proof that international broadcasting will remain the main gateway for the projection of foreign policy for a few more decades to come.

Chapter 2
An analysis of SW sales 1950–1997

Wars, whether hot or cold, act as a spur to the advancement of science and technology. International broadcasting in the high frequency spectrum, from 3.9 to 26.1 MHz, and the technology of the SW transmitter, owe much to World War II and the cold war. The growth in SW has been explosive.

Transmitters have also become more powerful. Output powers rose from 10 kW in the 1930s to 250 kW by 1965 and to 500 kW by 1985. At the end of World War II, Britain emerged as the most powerful propaganda broadcaster. Its weekly output in languages and programme hours exceeded the combined total of the two world superpowers, America and the Soviet Union. However, the onset of the cold war and the conflict of ideologies between East and West changed this. It fuelled the build-up of America's broadcasting arms, Voice of America (VOA) and Radio Free Europe/ Radio Liberty (RFE/RL).

From 1946 it was the US broadcast industry that dominated world markets with such well known brand names as Continental Electronics, Collins Radio, General Electric (GE), Hughes and Gates. This almost total domination of world markets was chiefly due to the fact that much of Europe's manufacturing capacity had been destroyed or badly disrupted. From the late 1960s European manufacturers were growing. The big electricals – Asea Brown Boveri (ABB), AEG Telefunken, Marconi, Siemens and Thomson – were all engaged in developing new and more powerful transmitters for the SW market. In 1972 ABB and Thomson CSF were the first to build a 500 kW transmitter using a single tube in the final RF stage. In 1978 AEG Telefunken took the technical revolution a step further with its Pulse Duration Modulation (PDM) technology. This was soon followed by another startling development, the Pulse Step Modulation (PSM) developed by ABB, which has since been adopted and improved by a number of other manufacturers [2].

5

The markets

The SW market is complex in nature and influenced by world politics. SW broadcasting is a political tool used mostly by governments and, because of this, contract awards tend to be influenced by factors other than the customary yardstick of cost and performance, plus other elements such as cost of ownership and reliability of service. It is well known that national interests, international politics and even commercial intrigue can and often do play a part in contract awards.

For example, no foreign bidder has ever been successful in winning a tender to supply French broadcasters with radio transmitters. The UK seems to adopt a similar policy of protecting British suppliers. The BBC has always bought from British suppliers, either STC or Marconi, but in the past few decades it has been Marconi with one exception (when it purchased four S4005 500 kW transmitters from AEG Telefunken in 1984). Yet the BBC never bought any more of the S4005, preferring the Marconi B6127 despite the fact that the former were more reliable and easier to handle and that their own engineers reported that the S4005s were of superior technology. Germany also has a long record of supporting German manufacturers; Deutsche Welle has never used any other transmitters than AEG Telefunken's, though with Deutsche Telekom Bundespost now privatised that policy might change.

The US Information Agency/Voice of America (USIA/VOA) procurement has a history of buying from different manufacturers at different times and this is how it came to have a relay network composed of more than a dozen different manufacturers' transmitters. However, the decision to award the VOA modernisation programme to GEC Marconi was judged by many to be wrong, as subsequent events seem to have proved. At the time it was felt that any one of four manufacturers could have been awarded the contract, Continental, Marconi, Telefunken and ABB. All had submitted models for type approval, and all were accepted as meeting requirements. Thomson had not submitted a model for type approval. The Marconi B6127 went on to win the contract. The VOA contract with GEC Marconi never went the way it was intended. In June 1997 GEC Marconi exited the high power broadcast market.

To their credit the other three manufacturers all recovered from any disappointment over the outcome of the VOA contract and forged ahead to introduce even more advanced SW transmitters, exemplified by the Continental 420C, the Telefunken S4105 and the ABB SK 55 C3 CPS.

Recent reports to come out of the VOA's Engineering and Technical Operations department indicate that they are highly pleased with the performance of Continental's 420C, 500 kW SW transmitter. This is upgraded from the model 420B by the retrofitting of Continental's all solid state modulator, resulting in a big increase in tube life in the final RF stage. TH558 tubes are now realising up to 30,000 hours. This is almost certainly the highest tube life ever recorded from a high power final RF amplifier stage. This is an important factor, because it means that the design of the RF stage, the most critical component in any SW transmitter, must be excellent.

Continental has been selected by the International Broadcasting Bureau (IBB) as the contractor to supply its solid state modulator to upgrade the entire SW transmitter networks of VOA and RFE/RL. A preliminary report indicates that the benefits gained are greater than was expected and the VOA is delighted with the results.

For many years I have been of the opinion that the US international broadcasting arms should, as a matter of national pride and in the tax-payer's interest, follow the policy of European broadcasters and support the national manufacturer. Certainly there is a good case because a company such as Continental, the only manufacturer of its kind in North America exists to serve the interests of its government. It is a strategic asset of great importance.

No other broadcast equipment manufacturer has served America and its people better than Continental Electronics. It has served successive US administrations for more than five decades. It has supplied high power transmitters for SW, MW and LW broadcasting. Its transmitters have performed in almost every region of the world, in a wide range of climate conditions and with a reliability seldom equalled, at the stations of VOA and RFE/RL.

In Europe, Telefunken has supplied Deutsche Welle and many other respected international broadcasters such as Austria ORF, Netherlands PTT, Norway NTA, Portugal RDP, Spain, the UK BBC, as well as Japan NHK, VOA and others all over the world. Altogether Telefunken has supplied more than 100 high power, SW transmitters. One of its largest ever contracts was for the supply of ten 500 kW SW type S4005 transmitters to the broadcasting authority of Iran. This was completed in 1995 and it also included two high power rotable curtain arrays and a considerable number of fixed curtain arrays. Telefunken has a reputation for quality of manufacture and reliability of performance, very rarely equalled but never bettered.

The SW market 1950–1990

From the 1950s the SW market saw a great expansion in sales. Though this market was fuelled by the cold war, there were two other contributing factors, increasing congestion in the SW bands, and continuing increase in man-made electrical noise, both of which had an adverse effect on signal audibility in the target zones. Paradoxically the introduction of more powerful transmitters brought more channel interference, thus bringing about a further spiral when major international broadcasters such as the BBC and Deutsche Welle began to introduce 500 kW transmitters into service.

Table 2.1 *Sales analysis for the four major European manufacturers [2]*

Period	ABB		Thomson		Marconi		AEG	
	[Units]	[kW]	[Units]	[kW]	[Units]	[kW]	[Units]	[kW]
1950–1954	4	400	1	250	4	400	1	50
1955–1959	7	700	2	600	3	300	–	–
1960–1964	13	2350	14	1530	27	4350	22	1950
1965–1969	18	4350	10	1350	26	6050	13	1300
1970–1974	29	7350	25	8150	11	1850	7	1450
1975–1979	19	6450	17	5350	11	3050	14	6750
1980–1984	47	14 750	27	8950	15	5300	16	6750
1985–1990	106	30 500	32	12 600	21	6750	26	10 200
Total	243	66 850	127	37 780	118	28 050	99	28 450

Units = Numbers of transmitters sold in five year period.
kW = Total power of transmitters sold.

Throughout the cold war period, from 1950 to 1990, total sales by the big four European manufacturers totalled more than 600 units with output powers between 100 and 500 kW, averaging at a little more than 120 units per five year perod. With 200 units sold in the last half decade 1985–1990, the average was 40 per year for that period.

Analysis of Table 2.1 shows ABB to be market leader with 243 units sold, Thomson-CSF second with 127 units sold, 118 units for Marconi and 99 for Telefunken.

The SW market in the post cold war period 1990–1993

This analysis embraces the post cold war period with the object of assessing any changes in sales. It also marks the last period for ABB, before its acquisition by Thomson, and the emergence of Continental Electronics as a major contender in the 500 kW market, with the first sale of its model 420C to the religious broadcaster EWTN.

Presented in a slightly different way to Table 2.1, Table 2.2 breaks down transmitter sales into three power levels – 100, 250–300 and 500 kW – with totals for the five manufacturers and for the overall number sold in each category.

Over the total period of four years the number of units sold was 156, an average of 39 per year. This is slightly below the average for the previous five years. As might be expected, ABB is market leader with about 38 per cent of the market in volume of sales and almost 40 per cent in terms of kW sold. Thomson is second with 23 and 21 per cent, respectively.

Table 2.2 *Cumulative sales of high power SW transmitters by the big five companies 1990–1993*

Company	100 kW	250–300 kW	500 kW	Total units sold	Total kW	% world market kW	% total units
ABB	12	22	28	62	20 700	38.0	39.7
Thomson	7	2	23	32	12 700	23.3	20.5
Marconi	0	11	23	34	12 600	23.2	21.8
Telefunken	0	0	10	10	5000	9.2	6.4
Continental	14	0	4	18	3400	6.3	11.5
Totals	33	35	88	156	54 400		

Units = Numbers of transmitters sold in five year period.
kW = Total power of transmitters sold.

The Marconi share of the market is artificially boosted by the VOA order. If this were discounted its performance would be unspectacular. Continental's share of the 500 kW sector is low because it did not enter as a major contender until 1993. However, in the 100 kW sector this company sold more transmitters than any of the other manufacturers, capturing over 42 per cent of the world market.

The SW market in the post cold war period 1991–1996

This period has been selected for analysis because it encompasses a number of important events; the ending of the cold war and the tearing down of the Berlin Wall and reunification of Germany, all of which might be expected to have had an effect on the SW market insofar as Western broadcasters are concerned. In the euphoria that followed these events, opinions were expressed by some in the broadcast world that international broadcasting in the HF spectrum had seen its finest hour in helping to bring about the collapse of communism, and that this might be the time to shut down some of the transmitters of RFE/RL and 'bring the boys home'.

Others took a more cautious view, pointing out that it could take as long as two decades to re-introduce free media to 200 million people who had experienced nothing but state control for six generations. This view-point saw RFE/RL in a different light, proof that a strong capability in international broadcasting should be retained by the United States, on the basis that no one can know what lies ahead in today's troubled world.

In the confusion that followed, some international broadcasters, such as Radio Canada International and Radio Australia, came in for budget cuts, blamed on worsening economies. On the other hand the UK Government

Table 2.3 *World sales for the major manufacturers 1990–1996*

	Units 100 kW	% world market	Units 250 kW	% world market	Units 500 kW	% world market	Totals	% world market
Thomcast	36	47	36	69	60	53	132	54
Continental	41	53	2	4	15	13	58	24
Marconi	0	–	14	27	24	21	38	16
Telefunken	0	–	0	–	15	13	15	6
Total	77		52		114		243	

saw no reason to cut back its broadcasting arm, and the BBC World Service can produce statistics to show that it actually increased its total world audience.

The following analysis for the period 1990–1996 registers total world sales for all regions by the big four manufacturers, Continental, GEC Marconi, Telefunken and Thomcast (ABB now having been absorbed with Thomson CSF to become Thomcast). Continental, which was not included in the 1950–1990 survey, was now in second place in terms of units sold.

The following notes of explanation are relevant here:

(1) Marconi sales are artifically boosted by the VOA modernisation programme, and all but two of the 500 kW class were built in America.
(2) The low sales by Telefunken are probably accounted for by the high level of the Deutsche mark and the effect of the high costs of the social programme following German re-unification. The restructure of Telefunken following its acquisition by Tech Sym is believed to have made Telefunken products more competitive.
(3) Thomson and ABB combined to form Thomcast, so figures reflect total combined sales of the companies. It will be noted that the total of 54 per cent of world markets represents a drop from 60 per cent in the earlier sales survey.
(4) The total value of the SW market is of the order of $300million. A 500 kW transmitter sells for just under $2million. However, this is only one component of the costs of a high power SW station.
(5) The emergence of Continental as a major player is obvious from an analysis of the market shares of the four companies. In the 100 kW sector it captured 53 per cent and across all power levels its share was 24 per cent.

Figure 2.1   *500 kW ABB SW transmitter. On the right the pulse-step-modulation and the transmitter control system, on the left the high frequency stages and in the centre the high power final stage transmitter valve*

Chapter 3
SW listening audiences and broadcasting output

Although a number of the government-funded international broadcasters such as Voice of America (VOA) and the BBC World Service (BBC-WS) collaborated to some extent in the cold war period, with regard to transmission schedules and programmes for their broadcasts to the Soviet republics and to those countries in Eastern Europe behind the iron curtain, it is a fact that outside these regions of the world VOA, the BBC-WS and indeed almost all of the international broadcasters competed for world listening audiences – often in the same regions of the world. For instance, both the VOA and the BBC have tried to increase audiences in parts of Africa and in SE Asia and, towards this end, both inaugurated plans to improve audibility.

The key to achieving bigger listening audiences, particularly in regions of the world where the broadcasters's ambitions may be viewed with a certain amount of distrust, rests upon two important criteria:

- The credibility of the programme as perceived by the listener;
- Audibility of the broadcast in the target zone or country.

Both are important, although broadcasters place differing emphasis upon them.

The reason for this has to do with the behaviour of SW listeners, especially those who are new listeners. New listeners tend to tune into the loudest signal in a broadcast band. If they like what they hear these listeners will probably look for that same station again. Few will have the time to listen to differing stations and make a comparison.

More broadcasters, increasing competition

Since the mid-1980s a new, powerful sector of the SW broadcasters has appeared, with a number of stations on the SW bands. These are the religious broadcasters and the development of this sector of broadcasting has increased the competition for SW listening audiences. These broadcasters are marketing faith. To the citizens targeted by them these broadcasters have a strong appeal. For many people the religious broadcasting station has an advantage over the international broadcasters. In a contest between faith and truth or logic, faith often appears to be the winner.

Christian and other religious broadcasters have a powerful advantage over other types of broadcasters as they usually deal in straightforward issues such as good versus evil, gospels or well refined tenets of faith. With their direct and possibly unambiguous messages, religious stations can polarise and win over considerable listening audiences, especially among the poor or oppressed, in the same way a missionary can be a more powerful force than a doctor. A doctor can mend broken limbs but the missionary can give hope for the future and the promise of a better life.

Simplicity of message appears to be a powerful advantage for religious broadcasters in terms of audience appeal, and since the mid 1980s this genre of international broadcasting seems to have won considerable audiences in Africa, China and SE Asia. However it is impossible to quantify the size of these audiences, or how much audience share these stations have won from other SW broadcasters, because these religious broadcasters tend to refer to potential listening audiences whereas the major international broadcasters produce listening audiences figures which are based upon detailed audience research carried out in the different regions of the world.

SW world-wide listening audiences

Some predict the demise of SW broadcasting based upon decline in world audiences. Such claims should be treated with caution and they should be analysed on a region by region basis. For example, if we examine the figures for regular SW listeners in Western Europe and North America, they show that a steady decline has taken place over the past five decades, and a realistic figure of regular SW listeners in these regions of the world might be as low as 25 to 30 people in every 1000.

The reasons for the decline are straightforward; they are to do with society which is undergoing change, becoming more pluralistic, with access to a wide range of media. SW has to compete in advanced societies with FM broadcasting, national AM broadcasting, television, video and Compact Disc (CD) and a broad and usually free press, plus increasingly the Internet. People have only so much leisure time, and for the financially better-off that time will be competed for by not just radio and television but also theatre, opera, ballet and popular music, to say nothing of sport which often dominates social activities.

It is also important to bear in mind that SW was never a mass medium in North America and Western Europe, so although there has been a decline in audiences, the decline has been a gradual one whose significance should not be exaggerated. In these regions SW listening was never a staple, and was never used as a medium for national broadcasting, so listeners generally fall into the sector of radio amateurs and DX SW listeners clubs, plus some travellers who like to travel with a pocket-type SW radio with which they can keep up with news and other information from their favoured international broadcasters.

At the opposite end of the social and income spectrum are the peoples of the Indian sub-continent, Africa and large parts of East, Central and South East Asia and South America. In many of the countries in these parts of the world circumstances are very different. Firstly, the short waves here are often used for national domestic broadcasting because it is the near-perfect medium for broadcasting over long distances. As a result the SW receiver is in common usage, listeners are familiar with the vagaries of the ionosphere and may well know where to find VOA and the BBC-WS on the dial.

Secondly, to the peoples of these media-starved countries, whose incomes on average may be a hundred times less than the citizens of North America, they do not have other competing forms of media and the SW receiver is often the only means of obtaining news of what is happening outside their own country. To these audience the SW receiver is not a luxury, it is a necessity, and unlike the citizens in the West who have a wide choice of entertainment, these people do not enjoy such choice. Figures for numbers of SW receivers in use throughout Africa, India and Asia are hard to assess on a country by country basis but may be as high as one in every three of the population. It is important to realise that SW listening to international broadcasts, as distinct to national broadcasts, has never been a mass medium but rather the means of obtaining foreign and sometimes domestic news. Hence, for both the originating broadcaster and the listener, the role of the SW broadcast has been to provide either an alternative or even sole source of news (albeit prone to the influences of propaganda or counter propaganda). In this sense the value of SW broadcasting was proven in the cold war, when for every one listener to the SW broadcasts into Eastern Europe and the USSR from VOA, BBC, DW, Kol Israel or RFE/RL – the most influential of the Western broadcasters in the minds of the Kremlin – the news was relayed by the word of mouth to another 10 citizens or more, notably in countries like Czechoslovakia.

International religious broadcasting stations like AWR, FEBC, FEBA and TWR have always appreciated the importance of their message being repeated by word of mouth. After all, this is how most religions began! Between these two extreme examples of SW audiences, the affluent Western societies with pluralistic media, and the more impoverished peoples in developing countries, there are two more categories of listener. One of these is the peoples of the former Soviet republics – of which Russia is by far the biggest – and the countries of the former Soviet bloc countries; the other is the English-speaking audiences to be found in almost every part of the world because English is, historically at least, the leading of the universal languages.

In the case of the former category the period between 1989 and 1995 saw some far-reaching changes in the listening habits of the peoples of the former republics of the USSR and the countries of Eastern Europe. During the dark days of the cold war, broadcasts from the Western broadcasters were not only a source of invaluable news on what was happening in other parts of the world, they were a source of inspiration, succour and hope to nearly 300 million people. But, more than that, being able to listen to the broadcasts of RFE/RL, whose transmissions penetrated the iron curtain for almost 24 hours per day, was a kind of a prestige status, and the information in these broadcasts, calculated by the broadcasters to be damaging to the leaders in the Kremlin, was passed on by word of mouth. As the cold war progressed, so the listening audience figures grew.

With the end of the cold war and the collapse of communism the political reality reduced the incentive to tune into RFE/RL, the BBC and VOA and audience figures began to fall, gradually at first. There were other factors hastening the decline in audience figures in this region too, including re-broadcasting agreements that the BBC, VOA and DW signed up with the former republics. Following the end of the cold war these three big international broadcasters concluded agreements with countries in Eastern Europe and various former Soviet republics to have their programmes re-broadcast over local FM stations and via some MW transmitters. Re-broadcasting usually gives a stable signal with better clarity, free from the vagaries of SW reception and to some extent broadcasters like VOA were victims of their own success. Re-broadcasting attracted a larger listening audience but SW as the means of transmission was the loser. Why listen to programmes from the West on SW if the same programme could be heard over the local FM station? In 1995 a listening audience survey carried out by VOA in the Czech Republic showed that of the total number of Czech listeners to VOA programmes, 64 per cent said they listened to VOA over their FM station.

Another factor which has played its part in the loss of SW listeners is the appetite of these listeners for new technology. In the days of the cold war AM broadcasting was the national staple and there were very few FM stations. With the wider introduction of FM broadcasting and technology such as the compact disc, listeners began to appreciate a better quality of sound. The short waves are suitable for listening to news programmes but there is no denying that for music SW is inferior. As more and more of the citizens of the former USSR get to appreciate better sound quality, it is likely that SW listening audiences in these countries will continue to fall off gradually.

If the reality of the end of the cold war and political reform was to diminish the appetite of listeners for the broadcasts of Western SW stations, the same can be said of the broadcasters themselves. There was and is no longer the need to provide a ceaseless bombardment of Western news round-the-clock to these formerly oppressed peoples, and barrage broadcasting, the technique where the same news was transmitted on many different frequencies and in several broadcast bands, has become a thing of the past. After all, whatever the excuses given by the Western broadcasters for these practices during the cold war, the fact remains that SW broadcasting is used for projecting foreign policy – and the policy objective had largely been met by the mid 1990s. In the uncertain world in which we live, with regional and border conflicts which can turn into a full-scale war, there is no shortage of new target zones to which broadcasters like the BBC World Service can direct their best efforts.

Surveys by the BBC have shown an increase in its listening audiences in some countries, though these are contradicted by another survey by VOA, which concludes that world audiences for SW are shrinking. What is a proven fact is that in any region more people tune into the short waves during a period of political tension, and that war especially gives a major boost to SW listening. Therein lies the true value and potential of SW broadcasting – it can reach out to people of all classes, so that even a poor farmer in a third world country can hear direct from studios in Washington DC, London or Paris, what is happening within his own country.

Universal language broadcasting

As a general rule, the language of a broadcast defines its target zone. For example, a broadcast in Croat would usually be directed to Croatia. However, in the case of the universal languages, of which English is the prime example, the target zone becomes the world. This brings us back to an advantage of the short waves and the fourth category of listener, English-speaking audiences. SW is the only form of media that can reach out to listeners in any part of the world without the intervention of a third party. Even a geostationary satellite system of three to four satellites could not cover arctic and antarctic regions, and the third party which owns the satellite(s) will always have the power of veto – sometimes called the gatekeeper effect – if it does not like what is being transmitted.

The international broadcaster which has a most positive view about SW is the BBC World Service. Some would say that with a world listening audience of over 140 million the BBC can afford to take such a healthy view. Certainly no other broadcaster comes close in performance. The BBC-WS does have the advantage that it broadcasts in the most widely spoken language. VOA does too but it achieves a lower world audience, despite the fact that it has greater transmission capacity. As might be expected, therefore, the BBC-WS has no plans to cut back on its SW and MW transmission capacity and it believes that classical SW broadcasting will continue to be the main gateway for direct international broadcasting for at least another fifteen years.

Technical drawbacks and scheduling

It would be wrong to suggest that the short waves are the ideal medium. All transmission systems have their shortcomings and it could be said that SW broadcasting has more deficiencies than other transmission methods, and its listeners put up with signal fading and surging, coupled with adjacent channel interference signals, to name a few problems. One of the biggest nuisances is that of electrical noise. Man-made noise has multiplied a million fold over the past five decades and is especially high in urban areas and large cities. Another serious problem is that unless the listener has an expensive receiver there is usually some difficulty tuning into a signal. Old age, coupled with arthritic fingers, often makes the task that much more difficult. All of these difficulties, together with the tendency for SW signals to be affected by astronomical factors over a period of seconds (due to changes in the ionosphere), mean that some listeners find it easier to settle for the strongest signal in a particular broadcast band, and if they like what they hear they may stay tuned to that station.

This is one of the reasons why SW broadcasting is highly competitive and it also accounts for the need for these international broadcasters to pay attention to ensuring good audibility in all their target zones by the use of high power SW transmitters coupled to highly directive curtain arrays.

An important factor for an international broadcaster seeking world audiences is to ensure that its broadcasts an be heard at prime times in any part of the world, and ideally at all times, even during the early hours when listening audiences are at the lowest. This will of course, entail the use of many powerful 500 kW transmitters at several strategic locations in conjunction with many directive curtains.

Conclusions

For the nations involved, is it all worth it? The answer is a definite yes – there is at present no better way for a nation to project its culture and foreign policy to the rest of the world. Many of the overseas listeners to the BBC-WS, for instance, are the elite – the decision-makers. As to the costs of operating an international SW service, the BBC-WS achieves a world audience which matches its annual operating budget of, approximately £140 million. This corresponds to one pound sterling each year, or two pence per week, per listener. No other form of media, to date, has been so economical.

Not every broadcaster can claim to achieve the same listening audience as the BBC-WS. In terms of listeners per kW of transmission power – which is a meaningful method for assessing efficiency – it is ahead of all its competitors, but even if we use some average performance figures they show SW broadcasting to be one of the cheapest forms of media.

In conclusion, if some international broadcasters are making cuts in their programme hours, or broadcasting the same programme on fewer frequencies, then perhaps we should look upon this as a good thing. This is because one of the most serious problems associated with international broadcasting is congestion in the airwaves. A recent survey shows more than 4000 channels assigned to SW stations, with the result that in some instances more than a dozen stations may be on the same frequency. Some of these stations fail to achieve good audibility in their intended target zone but somehow, due to the vagaries of the ionosphere or by incorrectly orientated antennas, manage to achieve good audibility in another part of the world, so making it difficult for a listener to hear the station they actually want.

In this context one of the best things that has happened since the ending of the cold war is that many of the world's international broadcasting authorities have for the first time in history collaborated. Those that have worked together include international broadcasters, religious broadcasters, surrogate broadcasters and even some who had operated unlicensed transmitters jammers and freedom stations. Such events would have been unthinkable in the early 1980s. In 1994 these former competitors on the SW bands got together for the first time to co-ordinate frequency and time schedules in order to reduce the possibility of interference.

These meetings, organised under the name 'High Frequency Coordination Committee' (HFCC) by the European Broadcasting Union (EBU), and attended by such leading SW broadcasters as VOA, the BBC, DW, RFI and RFE/RL, as well as Radio Moscow and some other broadcasters from Eastern Europe, were without question groundbreaking events in SW international broadcasting, and represent the best efforts yet for reducing interference (defined as 'collisions' by the HFCC).

Broadcasting output from the world's international broadcasters

The period from the ending of the cold war and the collapse of Soviet style communism has been followed by a period of uncertainty not only in the West but throughout other regions of the world. Some see the world as a safer place, whilst other political analysts see a world which has had its balance of power radically shifted with the result that uncertainty prevails.

Some international broadcasters have taken the view that SW international broadcasting, having made its greatest achievement in playing a pivotal role in the ending of the cold war, should now be wound down to a lesser role. Some others such as the BBC-WS and VOA believe this is no time for a nation to give up its international broadcasting capability.

Some of the major international broadcasters have cut back on pro- gramme output since 1991, and one or two have been threatened with closure or have experienced cutbacks in operating budgets. If an inter- national broadcaster cuts back on programme hours or transmission capa- city, then a fall in listening audiences will follow automatically, thus giving some analysts further evidence in which to conclude that SW listening audi- ences are in decline, an obvious conclusion but not necessarily the correct one. This is borne out by world audience statistics for international SW broadcasters. The BBC-WS, is universally accepted as the leading broad- caster in terms of its 140 million listening audience, though it is not the world's largest in its transmission capacity as it is exceeded by VOA (whose audience is nearer 105 million).

Moreover the BBC-WS steadily increased its share of the world listening audience with each passing year over the past few decades, right up to the collapse of Soviet communism and the tearing down of the Berlin Wall. This performance by the BBC-WS was not surprising, but what has been aston- ishing is the way this same international broadcaster has actually gone on to increase its world listening audiences after the end of the cold war, while some broadcasters subsequently lost listeners by cutting back on programme hours.

The BBC-WS increased its world audience from 120 million by 1992 to 130 million by 1994, to 133 million by 1995 and 143 million at the end of 1998. What these audited figures from IBAR (International Broadcast Audience Research) show are firstly that the BBC-WS is providing a service which meets with the approval of its audience, and, secondly, that BBC SW listening audiences measured on a global basis are not decreasing but increasing.

Broken down, the results from the BBC-WS IBAR show large variations in SW audience size from one world region to another (Table 3.1).

Table 3.1 *BBC-WS regional audience figures 1995 (direct and indirect)*

Region	Audience (millions)
Africa and the Middle East	48
Americas: North, Central and South	9
Asia: Pacific region	8
Europe: West to East	14.5
FSU and SW Asia	8.5
South Asia	51.5
Total	139.5

Table 3.2 *Estimated total direct programme hours per week of some external radio broadcasters*

	1950	1960	1970	1980	1990	1996
United States of America	497	1495	1907	1901	2611	1821
Chinese People's Republic	66	687	1267	1350	1515	1620
United Kingdom (BBC)	643	589	723	719	796	1036
Russia	533	1015	1908	2094	1876	726
German Federal Republic	0	315	779	804	848	655
Egypt	0	301	540	546	605	604
Iran	12	24	155	175	400	575
India	116	157	271	389	456	500
Japan	0	203	259	259	343	468
France	198	326	200	125	379	459
Netherlands	127	178	335	289	323	392
Israel	0	91	158	210	253	365
Turkey	40	77	88	199	322	364
North Korea	0	159	330	597	534	364
Bulgaria	30	117	164	236	320	338
Australia	181	257	350	333	330	307
Albania	26	63	487	560	451	303
Romania	30	159	185	198	199	298
Spain	68	202	251	239	403	270
Portugal	46	133	295	214	203	226
Cuba	0	0	320	424	352	203
Italy	170	205	165	169	181	203
Canada	85	80	98	134	195	175
Poland	131	232	334	337	292	171
South Africa	0	63	150	183	156	159
Sweden	28	114	140	155	167	149
Hungary	76	120	105	127	102	144
Czech Republic	119	196	202	255	131	131
Nigeria	0	0	62	170	120	127
Yugoslavia	80	70	76	72	96	68

Notes:
(1) USA includes VOA (992 hours per week), RFE/RL (667 hpw), Radio Marti (162 hpw) – 1996 figures.
(2) Since the break-up of the former USSR in 1991, only Russia's output is shown.
(3) 1996 figure for Czech Republic (created 1.1.93), previous years for former Czechoslovakia.
(4) At the time of going to press, South Africa's external service's future is in doubt, and Nigeria's external service is off air.
(5) The list includes about a quarter of the world's external broadcasters whose output is both publicly funded and worldwide. Among those excluded are Taiwan, Vietnam, South Korea and various international commercial and religious stations.
(6) 1996 figures as at June; all other years as at December.

Source: International Broadcast Audience Research, June 1996.

Programme output hours

The most widely used parameter in international broadcasting is that of programme hours output, usually measured on a weekly basis. Such figures are a direct sum of all language programmes radiated by an international broadcaster by SW broadcasts (and by MW where used). Such figures do not include re-broadcasting. Thus one transmitter would be capable of transmitting 168 hours a week if used continuously. In practice, most broadcasters like to reach audiences at prime time, when numbers are at their highest, as broadcasting in the small hours is less cost-effective. In the cold war there was a notable exception in that RFE/RL ran its transmitters round-the-clock. This was because many of the regular listeners were of a more 'covert' type.

   Programme hours output is a useful guide to the relative size of the international broadcasters. The most reliable guide to programme hours is published by the BBC's IBAR department. Because this publication shows figures for each half-decade back to 1950 it is possible to identify trends. Table 3.2 shows such figures for most of the major international broadcasters back to 1950. Those whose figures are not published by IBAR include Taiwan, Vietnam, South Korea and various international commercial and religious stations. However I have obtained some figures for June 1996: Taiwan 925 hours, South Korea 350 hours and Vietnam 210 hours. Figures for the leading international religious broadcasters are given in Chapter 21.

Chapter 4
Projecting foreign policy, propaganda, beliefs and objectives

Today the dissemination of news and information has become a major force in the world, almost ranking alongside that of capital and labour in its importance, and as the Western societies move towards a mix of service-based and leisure-based industries so its importance will grow. Information has become the critical currency. Yet outside North America and Western Europe there are many countries which have never known what it is like to have a free media, and have only recently begun to move in that direction. These are the countries of Eastern Europe and the former Soviet Union (FSU).

But beyond the former Soviet empire there are the peoples of Asia and large parts of Africa, many of whom live in poverty, or under oppressive or unstable regimes who never get to know what is going on in their own country or the rest of the world. Fortunately, it so happens that in many parts of Africa and Asia such as Zaire, Malaysia, Indonesia and FSU, SW is used by national broadcasters and therefore the standard portable radio receiver with a SW facility is often the only means of hearing news, either from domestic broadcasters or the rest of the world. For these peoples listening to international broadcasters such as the BBC World Service (BBC-WS), Voice of America (VOA) and a handful of other Western broadcasters the short waves are a lifeline.

Most international broadcasters take pride in the fact that they are the voice of their country, exemplified by station identification, Voice of America, Voice of Turkey etc. There are some exceptions; for example the BBC-WS, which goes to some lengths to claim the opposite – that it is not the voice of Great Britain but the voice of the independent BBC. For the most part these international broadcasters reach out to world audiences in a number of different languages and although there is a mixture of programme content, news is usually the main element.

Today some 800–900 million people tune into these international stations and the major broadcasters command large listening audiences, and by virtue of this fact are able to play powerful roles in the world. Though the major Western broadcasters do have a high degree of editorial independence in the type of programmes broadcast, listeners need to remind themselves that by the terms of their charter or licence the broadcasters are usually required to operate in the interests of the state. For this and other reasons internationally broadcast news is seldom neutral. As with most news, its balance and objectivity depends on its source, its editing and presentation, and factors such as emotion, logic, carefully sifted rhetoric and elaboration. Differences between objective reporting and political bias are not always easy to detect. Neither is it necessary to depart from the truth if there is a political motive, but rather to leave something unsaid.

During the cold war Western news agencies and broadcasters cooperated with Western governments to acquaint the West with life under Soviet communism. What was reported about the secret police, the KGB, lack of freedom and the state-controlled media was all true. What was not told were the fine features of Soviet society, job security, social services and a better balance of society with more equal distribution of wealth. Western perceptions of Soviet life were carefully shaped to project an image of a failing or inefficient society.

At all times international broadcasters have the difficult task of riding a fine line between retaining editorial freedom to report honest news and comment, whilst remembering they have a responsibility to broadcast in the interests of the government of the day. For non-politically aligned countries such as Sweden, Norway and Switzerland this was never a problem, but for politically aligned countries like Great Britain and the Federal Republic of Germany it was a different matter. In the context of politically aligned countries during the cold war, a former KGB agent confided 'In the cold war we had to constantly monitor and analyse broadcasts from the BBC World Service. It was one of the Western broadcasters we feared most. Whilst the Soviet system was teaching its peoples education, i.e., how to think, the BBC was teaching the listeners what to think – but were doing it in a subtle, refined way that was not easy to detect.'

Taken at its broadcast level, information is seldom neutral, especially that which has a social or political import. Either by choice of what material and views are to be presented, or by the way such material is presented, perceptions are shaped in the minds of the listeners. Expanded news coverage on international events tends to be governed by political responsibility by the broadcaster. This is where a broadcaster presents extreme views or partisan views which are in the interests of the state. Nevertheless for it to receive an audience, news has at all times to be credible. Clause 2 of the BBC-WS objectives states that news must be credible, whilst clause 3(b) states 'International news takes in a British balanced view of these developments and of world events in general, taking into account British Government Policy.'

Propaganda

Since the word propaganda crops up from time to time in this volume, and usually in the context of international broadcasting, it seems appropriate to attempt to dispose of some of the perceptions of the word itself, associating it with lies and deception. Such perceptions exist in Great Britain. Staff at the BBC-WS look over their shoulders when on occasion I have brought the word propaganda into conversation. 'Please,' I was once implored, 'do not use that word within this building.'

The BBC-WS is not the only establishment in Britain to feel ill at ease with the word when it is used to describe the role of the service; the Foreign Office is another. The average Member of Parliament is under no illusions about the meaning of the word; 'That's what the foreign countries do' one MP told me. This same attitude can even be found amongst senior members of government. Yet beyond Britain itself and into Western Europe, a different meaning prevails. No eyebrows were raised at Deutsche Welle or at Radio France International in Paris when I brought the subject of propaganda in international broadcasting up in conversation.

The best example of differing perceptions of the meaning of 'propaganda' in Great Britain and Germany was in 1940. In the Houses of Parliament some MPs demanded that German propaganda broadcasting to England should be answered with 'truth' from the BBC in London. But at about the same time in Berlin, Dr Joseph Goebbels, the Reich Minister for Propaganda, was admonishing a Berlin news agency for saying that the BBC was broadcasting propaganda to the German people. 'That is untrue,' Goebbels said, adding 'the word propaganda must only be used to describe German broadcasts. The BBC is not broadcasting propaganda, it is broadcasting lies.'

Goebbels was a classical scholar with a deep understanding of European history. He made propaganda his passion and mission and the success that he enjoyed as a propagandist came from his sixteenth century studies. That was a period when the catholic church was having its authority undermined by the counter-teachings of Martin Luther, the German reformist. Faced with the loss of whole nations from the catholic faith to protestantism, Pope Gregory XIII (1572–1585) sought a remedy in the building and endowing of colleges and seminaries under the authority of the Sacre Congretario de Propaganda Fide. This powerful body was given responsibility for the re-building of the catholic church. But if Goebbels was impressed with the workings of this propaganda ministry of the catholic church he was also attracted by the way Martin Luther had propagated his own beliefs throughout the German states and to other countries in Europe.

Luther's Reformist Movement was religion in partnership with the state and it was through Martin Luther and his reformist church that the different states of Germany eventually became a united Germany. This was one more fact which no doubt made Goebbels even more aware of the power of propaganda. Subsequently another Pope, Gregory XV, who though only Pope from 1621 to 1623, sought to arrest the spread of the Lutheran movement by passing a special decree for establishing a permanent congregation for the control of foreign propaganda ministries, whose duties were to project the faith.

From these early origins propaganda can be defined as a one-way communication system designed to influence belief. In its classical form propaganda is a moral uplifter and a sustainer. However for it to be successful there are certain key rules which must be followed. It was Pope Gregory XV who defined the first rule when he said 'The masses have a great capacity for forgetfulness, therefore the message of God should be simple, and it must be repeated over and over again.' Since then, others have been adopted but one of the most important rules to be followed is that it is first necessary to gain the trust of the people.

Although propaganda had its origins in the sixteenth century, another two centuries were to pass before governments saw it as a tool to influence people. This was due to the fact that newspapers had not reached mass circulation figures. The first attempt by any government to make use of negative propaganda came towards the end of World War I. In 1918 the British Government set up an organisation to undertake the task of disseminating propaganda to both enemy and neutral countries. From Crewe House under Lord Northcliffe there soon emerged a steady flow of propaganda. This was in the form of leaflets delivered by air to German troops in the trenches and intended to demoralise. The contents of these leaflets were soon communicated to the civilian population in Germany by troops on leave. The Crewe House propaganda unit is acknowledged to have played a key role in the unexpected breakdown of Germany's ability to continue with the war.

The advent of mass circulation newspapers was followed in 1920 by the introduction of national radio broadcasting in many European countries, which gave further opportunities for governments to influence people's thinking. World War II then gave the first opportunity for corruptive propaganda broadcasting to be used as a weapon of war. The effectiveness of such, sometimes called black broadcasting or negative propaganda, should not be underestimated. Yet the achievements of wartime propaganda cannot be compared to the success of broadcasts that promoted German National Socialism. In the 1930s Adolf Hitler perceived that for his political party to be successful its beliefs had to be propagated by every possible means, to the end that it might become the faith of the German people. Hitler and Goebbels had a role model to use from contemporary history: the development of communism, which the Soviet Union had adopted two decades earlier.

From the 1940s the allies began to realise that propaganda was capable of converting political dogma to a faith for a nation to follow. That realisation was reinforced at the end of World War II when the victorious Allied powers were faced with another task; that of how to de-Nazify a nation. The lessons learned by the Allies in using increasingly sophisticated broadcasting were put to good use in the cold war with the Soviet Union which followed from 1949. On the premise that a faith, belief or a religion cannot easily be disposed of by logic, but best by the substitution of another belief, the Western powers enlisted the aid of powerful international religious broadcasters to project Christianity and the Zionist faith to such ethnic groups in the USSR during the cold war. Kol Israel (Voice of Israel), the international broadcaster of the Jewish State, broadcast to the ethnic communities in the USSR and Eastern Europe in Hebrew, Georgian, Bucharian, Hungarian, Ladino, Mograbit, Romanian Russian and in the Yiddish tongue.

Religious propaganda broadcasting is also playing a key role in another political confrontation. This is the long-running 'near-cold' war between the United States and the People's Republic of China. Like Soviet communism, Chinese communism is a belief but with a much bigger following – approaching 1.2 billion. In one of those remarkable coincidences that seems to be a feature of American politics, the Far East Broadcasting Company, a religious broadcaster registered in California in 1945, was broadcasting the Christian faith to communist China in 1949 shortly after that country had declared itself the People's Republic. Currently communist China is being targeted by a number of US broadcasters from Pacific bases.

Propagating a faith or a belief

There are important rules to be followed if propaganda is to be successful, and techniques have been honed and refined over time. Some of these are:

- to first gain the trust of the people;
- to lay claim to being an elite movement or body;
- to project total credibility in its claims;
- to make it possible for anyone to become a member.

Others include brevity, repetition of key words, simplicity and the liberal use of symbols or slogans. Political parties of all kinds and nationalities are principal users of propaganda and have developed these techniques. Propaganda is intended to appeal to audiences in the following ways:

- it promotes a warm and pleasant feeling of belonging;
- it promotes ideas, thoughts or opinions that seem reasonable;
- it may suggest a course of action that seems justified.

There is no exact definition as to what constitutes propaganda. However the closest is to compare it with education. Propaganda is the antithesis of education. Whereas the educationalist promotes objective thought and original thought and seeks consensus, the job of the propagandist is to present a particular viewpoint, or in the case of a religion to promote a belief, to suppress analytical thought and to reject consensus. In brief the propagandist seeks to build up the strongest possible case for its message.

In the context of the government-sponsored or government-funded international broadcasters, with which this volume is primarily concerned, these media represent their countries in the international broadcasting bands and thus broadcast in the interests of the state – or are required to take foreign policy interests of their respective governments into account.

History shows that societies are capable of throwing up individuals from time to time who, through force of character, coupled with personality and orchestrated propaganda, often based upon personal conviction, can change the character of a generation of people. Such people have included Senator McCarthy of the 1960s in America who instituted a campaign of fear based upon his hatred of communism, and some would argue Margaret Thatcher, former Prime Minister of Great Britain, who in the 1980s changed the character of a generation and almost succeeded in disrupting the future of Europe for ever as a result of her disagreement with Germany and France.

But such consequences cannot be compared with those of despots and tyrants who have spawned evil regimes in parts of Africa, Asia and elsewhere in the world. This is why the major international broadcasters like the BBC-WS, VOA and others continue to reach out to oppressed peoples wherever they may be in the world, and hope to sow the seeds of democracy and a better life. Unfortunately the evidence seems to suggest that those countries most targeted are left wing or communist countries, and do not include repressive right wing countries.

Censorship of foreign broadcasting

In November 1995, the British government through its Department of National Heritage imposed a nationwide censorship on incoming TV transmissions from an uplink satellite programme from Sweden broadcast from two European satellites, EUTELSAT 11 F3 and 'HOT BIRD 1'. These powerful direct broadcast satellites cover the whole of Europe. The programme which the Department of National Heritage censored was a Swedish pornographic programme for adult viewing called 'XX TV Erotica'. The government could not actually prevent viewers from tuning into the transponder frequency, but it achieved total censorship by making it a criminal offence for the UK-based company Media Satellite to market and sell the necessary smart cards required to unlock the encrypted transmissions, and went one stage further by making it a criminal offence for the British press and media to advertise the programme.

This was not the first time the British government used draconian powers to censor foreign broadcasts as it did something similar in May 1993 with a pornographic satellite programme called 'Red Hot Television'. By both of these actions the British government seemed to be emulating those countries in the Middle East and the Gulf region where broadcasts from nearby countries are banned for political reasons. The British are in no position to claim the moral high ground when it comes to radio and TV broadcasting. Commercial broadcasting can be damaging, as it can influence the public to smoke cigarettes harmful to health, or to buy products they cannot afford. It is also conceivable that some adults might have their moral values corrupted by watching erotic films. Yet these consequences are as nought when compared to the possible consequences from the acceptance of certain religious or political dogma which runs counter to those of the countries in which listeners reside. In many countries the consequences or penalty for listening to certain foreign broadcasts was and is severe. For example, at best a SW listener in the Soviet Union who listened to the BBC-WS broadcasts during the cold war would be branded as a dissident, at worst he would have been banished to a concentration camp in Siberia for corrective treatment, from which he might never have returned. Yet, the BBC-WS and some other Western broadcasters continued to target listeners in the Soviet Union and Eastern Europe with broadcasts that contained material which ran counter to the teachings of the Soviet political indoctrine and which were intended to undermine those beliefs amongst ordinary citizens, and which were a real danger to a citizen if he or she absorbed the content of the broadcasts.

None of this is reflection on the Department of National Heritage, nor is it a criticism of the BBC-WS which is required by the terms of its royal charter to broadcast in the interests of the British state and to project foreign policy in its programmes. What this example shows is the duplicity and perfidity which all governments practise from time to time in the process of government. The two British government departments concerned with this story are the Foreign and Commonwealth Office (FCO), which determines foreign policy, and the Department of National Heritage which concerns itself with a wide range of domestic issues.

Thus a situation existed where one government department took steps to block broadcasts from another country because it thought its citizens should be protected from pornographic broadcasts, whilst, on the other hand, another department of government had used the BBC-WS to broadcast political propaganda to the citizens of the USSR, over the heads of the established regimes of the Soviet bloc countries, for nearly forty years, an action which posed a far greater threat to those listeners.

Chapter 5
Structure of US international broadcasting

The structure of America's international broadcasting operations is quite different to that of practically every other country. Unlike Great Britain which, officially at any rate, has only one foreign service broadcaster – the BBC World Service (BBC-WS) – the United States of America has a fairly elaborate structure of separate broadcasting organisations, some of which are government-funded and some privately funded. Yet each broadcaster has a duty to broadcast in the interests of the state and to play its own part in the business of projecting American culture, the American way of life and American politics to the rest of the world. In other words, to make the rest of the world 'user friendly' to America.

Foremost amongst America's international broadcasters is Voice of America (VOA), which is the international broadcasting arm of the US Information Agency (USIA). Its mission is to represent America on the international airwaves, to broadcast high quality news, views and some entertainment programmes to listening audiences around the world, and to articulate US government policy whilst preserving its editorial independence. VOA is funded directly by American taxpayers and is therefore fully accountable to the American public. By accepted definition VOA is a propaganda broadcaster and hence cannot go against the foreign policy of the government of the day. VOA is proud to be the voice of America.

Radio Free Europe and Radio Liberty (RFE and RL) are essentially quite different from VOA. Founded in 1950 and 1952, respectively, they functioned together to reach listening audiences behind the iron curtain and in the republics of the Soviet Union. Both of these broadcasters, called the 'voices' by the KGB, had an entirely different mission to VOA. Theirs was not to project the all-American way of life, but to appear to listeners as East European radio stations. To enable RFE and RL to fulfill this identity the studio staffs, producers, readers and news correspondents were emigrés – dissidents from the USSR and from the countries behind the iron curtain: East Berlin, East Germany, Hungary, Czechoslovakia, Poland, Bulgaria, Romania and Albania. In essence RFE and RL were seen as East European

29

radio stations operated by dissident voices from within the Soviet Empire. Their funding was on a different line to that of the VOA. RFE and RL were registered as one company with headquarters in Munich, privately owned and funded from charitable donations, though in actual fact the funding came from the US government through indirect channels from the US Board of International Broadcasting, whereas VOA was funded directly through the Bureau of Broadcasting and the USIA.

The third element to make the US a powerful international broadcaster in the SW bands are the privately owned, commercial religious broad- casters. Despite their private ownership, a condition of the US Federal Communications commission (FCC) terms of licensing is that there should be a need for their broadcasts to foreign countries, and that they broadcast in the interests of the USA. These private radio stations are able to play what some believe to be a powerful role in projecting American influence, with emphasis placed upon American democracy to regions of the world that are crucial to America's global interests.

The religious broadcasters are in a class of their own, using the tool of Christian religion to win large listening audiences in many parts of the world, especially in parts of Africa and Asia. The funding for many of the religious international broadcasters is said to come from religious donations, legacies and from the programme suppliers. This last-named source of funding seems to be at odds with the usual commercial practice where the broadcaster pays the programme supplier. The fact that it is a requirement for a religious broadcaster, even a private one, to carry programmes for, or on behalf of, another body lends itself to speculation and all kind of pos- sibilities. For instance, the US President's Task Force Document (1991), commenting upon the desirability of establishing a Free Radio for Asia to carry out surrogate broadcasting to the communist countries of China, Vietnam, North Korea, Laos and Cambodia, concluded that such a service should start up 'with a minimum delay using some existing transmitters which might be facilities of private US broadcasters.' It also goes on to say that surrogate broadcasting to Asia should be placed under the Inter- national Broadcasting Board, and that surrogate broadcasters should seek to broaden their base of support through underwriting by private enterprise or foundations.

Another US international broadcaster is the Armed Forces Radio and Television Service (AFRTS), or Armed Forces Radio Service (AFRS) as it was called during World War II when it started broadcasting in 1942 during the dark years when America was losing the war with Japan. AFRS began life with a few low power radio stations which covered small theatres of military operations. But by the time the war with Nazi Germany was in its closing phases AFRS had expanded to take over most of Germany's powerful radio stations (the Reich Rundfunk), some with 200 kW of carrier power. Today, this armed forces broadcasting network has radio and television stations located in 18 host countries and 'outlets' such as unmanned re-broadcast repeaters or unmanned cable operations in about 132 countries.

Though the role of AFRTS is to broadcast radio and TV programmes to American forces around the world, in doing so it is also projecting American culture, the American way of life and American products.

US national broadcasting

All national broadcasting in the United States is regulated by the Federal Communications Commission (FCC). More than 11,000 radio broadcasting stations, mostly private and commercial, were licensed to operate in 1995. For the most part these use low power (50 kW or less) transmitters. The interests of all US national broadcasters are served by the National Association of Broadcasters (NAB). The USA has a number of major broadcasting networks to which many of the private stations are affiliated and receive syndicated news and programme services. These major broadcasting networks are: ABC Radio Network, CBS Radio Network, CNN Radio Network, National Public Radio, Unistar Radio Network, and Westwood One Radio Network.

US national domestic broadcasters, whether private, commercial or religious, are not permitted to carry out international broadcasting. However, the US government can call upon the services of national broadcasters in times of disasters, national crisis or war.

Propagating American values

Successive American administrations from the very early years of the cold war have sought to justify to American people the nation's involvement in the cold war with communism.

The Third Reich was unique in its ability to mobilise an entire nation. It used a positive form of propaganda in its classical role as an uplifter of moral virtues and national pride. The USA took positive propaganda to another dimension by bringing God into the arena. From the early 1950s communism had been frequently equated with atheism. From this position it was easy to move on to the next phase and anti-communism became a kind of moral crusade in 1950s America. The cold war according to J F Dulles was a war against atheism. Moral and religious rhetoric was used in press statements from Washington DC. By the mid 1950s anti-communism had become the state policy. America declared itself to be the capital of the free world. In response to the rhetoric coming from Dulles and Senator McCarthy, a writer from Indiana told Dulles, 'God is trusting you to help redeem the world'.

There is an accepted view of Americans on the whole having moral and religious qualities coupled with a sure and unwavering belief in the American way of life, its constitution, and what the US can do for the rest of the world. Successive American administrations have nurtured these qualities.

With the cold war entering its second decade, the Kennedy adminis-tration decided it was time to explain more about how the cold war was pro-gressing, the role of America and how America would roll back the frontiers of communism, using VOA, to the American public.

The truth was something to the contrary. If any rolling back was going to be done, it would be by RFE and RL, but the existence of these two broadcasting agencies was not in the public domain. Accordingly, lectures and talks were given in US factories to tell American citizens how the efforts of many were directed to the war against communism, by peaceful means from the VOA.

One such talk was delivered at the Kiwi Club in Quincy, Illinois on 4th December 1961, the Mid-American city which is home to the Harris Broadcast Transmitter Division. Lawrence Cervon, then Vice President/General Manager at Harris, delivered his talk to the citizens of Quincy. He began by recalling the State of the Union Address to Congress, given by President Kennedy on January 30th, 1961:

> I speak before you in an hour of national peril. Before my term is ended we shall have to test anew whether a nation, organised and governed as ours, can endure. The outcome is by no means certain, the answers are by no means clear. All of us, together, this administration, this Congress, this nation must forge these answers.

That address by Kennedy, repeated at Quincy, fulfilled the basic rules of propaganda. It made an appeal to the people, it laid claim that the battle of the cold war was a just and noble pursuit, it gave to the listener a warm and pleasant feeling of belonging, and it was a unifier of the people. Cervon then went on:

> To all of us here, and to the nation, one of our greatest challenges today is the battle for men's minds. It is in this area where radio broadcasting, and Quincy, have been playing an important part. The products built here are vital to America and the free world. You see, as a weapon of war, radio is very important, especially shortwave. The activities of our government in the use of radio to tell the world about our aspirations for harmony, peace and good have never been given wide publicity. I'd like to tell part of that story, not only because it is important, but because Quincy

industry and people are playing a very important part in this activity.

Cervon then moved on to discuss the origins of broadcasting, SW, some facts about VOA, and how it could effectively combat ignorance and influence minds before saying that 'Radio Moscow allots more each year to jam VOA, than its entire budget.'

His talk was finished off with this message and testimony to the people of Quincy:

> You should be proud that Quincy industry has consistently been supplying American transmitters to help expand the voices of many nations that are facing the perils from communism. An important part confronting the world today is to provide radio communication facilities to the new nations of the world. It can combat ignorance, superstition and intolerance. As you have seen, the Voice of America is using transmitters built here in Quincy, a very important weapon in the cold war. In the days of the nuclear deterrent, many feel the ultimate contest between freedom and communism is the battle for men's minds. If we believe in our own importance and greatness, others will believe it too. Quincy is vital to our nation and to the free world, and all of us are vital to Quincy.'

Few were more ably equipped to lead, direct and to motivate workers at a high technology production plant than Lawrence Cervon. Under his leadership, first at Gates and then with Harris, the Quincy plant went on to develop and produce large quantities of equipment for military communications and radio broadcasting (see Chapter 23 for a history of Gates). In 1991 Cervon's efforts were recognised when he was honoured by the NAB for his 45 years of vision and leadership.

Similar talks were being given in other towns and cities where production plants existed. It was a way of uniting employees against a common enemy just like during World War II. Messages were designed to appeal to a person's moral values – to make them committed to the cause, if they were not already committed. Words were carefully chosen so as to convince people that they were taking part in a moral crusade, in this case the evil of communism. Cervon's talk was tailored to suit the moral virtues and a belief in the American way of life held by Mid-Americans. It had the right qualities and managed to convey the message that they were helping to build the weapons for the crusade to rid the world of the evils of communism.

Thirty years after Cervon had delivered his talk to the citizens of Quincy, and two years after the US had won the battle against Soviet communism, pronouncements were being made on how America would continue the fight against the evil of communism in Asia in the US Task Force on government international broadcasting.

US Government international broadcasting

On 29th April 1991 the White House, through the Office of the Press Secretary, issued a statement on the establishment of a task force by the President. This independent, bipartisan task force would study and advise on the best organisation and structure for US government broadcasting. In the light of dramatic political developments worldwide, which included the democratic revolution in Eastern Europe and the ending of the cold war, followed by the conflict in the Middle East, it was appropriate and timely to examine US Government international broadcasting.

The task force was required to make recommendations to the President within six months, on the overall context of US foreign policy and public diplomacy in the following areas:

- The most appropriate organisation and structure under which all US government international broadcasting assets would eventually be consolidated, in steps and over a period of time, under a single government broadcasting entity; and when and how such consolidation should take place.
- New technologies in light of the need for US Government broadcasting to remain effective and competitive. These were to include strategies for the best use of new technologies.
- The relationship between US Government broadcasting and US private enterprises in the international arena.

A committee of eleven members led by John Hughes, Director of the International Media Studies Programme, headed the task force and was made up of distinguished US citizens. It furnished its report to the President in December 1991. Comprising a main section with an eight-part appendix, the overall report ran to 98 pages. It is probably the most detailed, exhaustive and searching analysis and comment ever published on the subject of government propaganda broadcasting, and government-sponsored (surrogate) international broadcasting and reflects great credit on the committee. It could stand as a role model for other governments to follow, but this is unlikely to happen. Few countries in the world have an appetite for candour and freedom of expression on the scale of the USA, enabled by its Freedom of Information Act. One cannot conceive of the British government, FCO and the BBC World Service issuing a document which sets out in most explicit detail the role of government international broadcasting as a primary tool for the projection of foreign policy, and how government-funded international broadcasting agencies played a pivotal role in the bringing about of the collapse of communism in the Soviet Union and in the Eastern bloc countries. There follow a few extracts from the 1991 report:

- 'The cold war was a contest of ideas, our side won. A stake has been driven through the heart of totalitarianism. Communist totalitarianism has been severely wounded, but it has not expired elsewhere.'
- 'Whilst change has taken place in Europe and the Soviet Union there remains a world that is fluid and dangerous. The Middle East remains riven with violence and extremism, swathes of Africa and Asia need truthful information as they claw themselves out of political bondage. The US needs the capacity to speak directly with friends as well as those nations that are hostile.'
- 'Five of the six remaining communist governments are in Asia, a third of the world's population live there. Only when these governments change will the world be essentially free of the oppressive communist system.'
- 'A majority of the Task Force agree on establishing a Radio Free Asia to communist countries: China, Vietnam, North Korea, Laos and Cambodia. Such a service should start up with a minimum delay. Creation of a surrogate radio for Asia would be shrewd, we will not only be on the right side, but the winning side.'
- 'History will record that the funds of VOA and the radios [RFE/RL], taxpayers' money, were amongst the most useful national security dollars spent in this century. They sent out words, not bullets. Ideas and not bombs, and they broke down a wall, and helped break up the Soviet Empire.'
- 'Most Americas feel today that we have something useful to offer to the world, most Americans understand that in the nature of a democratic society it would be a shame if we did not offer what we have. Such a course of action is right morally, and from a point of view of self-interest Americans want a world that is user-friendly to our values.'
- 'There will be an indefinite and expanding mission for the Voice of America. Radio Free Europe and Radio Liberty should have continuing albeit modified missions.'
- 'The US should continue to make investment in shortwave facilities to the end of this century. US government international broadcasters should seek opportunity to place their programmes on foreign listening audiences so as to maximise world listening audiences.'
- 'The Task Force believes its [VOA] reporting has been first rate. Yes, VOA is the government-financed radio of a superpower. By congressional mandate VOA is obliged to articulate the foreign policy viewpoint of the US administration of the day.'

- 'New shortwave transmitters should have single sideband capability.'
- 'The International Broadcasting Bureau (IBB) should manage all US government-sponsored broadcasting. The Office of Cuban Broadcasting should be transferred to the IBB.'

Chapter 6
The BBC World Service

More than 150 countries use the international broadcasting bands in the HF spectrum. Almost all are government-funded and subject to direction and control in varying measure, depending upon the country. With some of these international broadcasters whose governments are not major players in global politics or are politically neutral, as in the case of Switzerland, government interference is less likely to arise. Nevertheless, by commonly accepted definition all can be described as propaganda broadcasters because according to the terms of their licence or charter they are usually required to broadcast programmes in the interests of the originating state. It is for this reason that most international broadcasters use titles including the name of their country, Voice of America, Voice of Turkey, Swiss Radio International and so on, and identify themselves as the voice of their country.

Not so in the case of the British Broadcasting Corporation. The BBC World Service (BBC-WS) goes to much trouble to claim the exact opposite: that it is not the voice of Britain but an independent voice of truth. Few of its listeners in India, Africa and other parts of the world would ever believe anything different and there are 130–140 million listeners who regularly tune into the BBC-WS in the SW bands. Yet, the plain fact is that the BBC-WS is funded by British government grants administered by the Foreign and Commonwealth Office (FCO), the government department to which the BBC-WS answers. The FCO works in conjunction with the BBC-WS to define target countries and hours of programmes. It would call for a large measure of naïvety for someone to believe that a truly independent broadcasting company would involve itself with global politics and play a pivotal role in the cold war, as the BBC-WS did.

Since it first began broadcasting to the British Empire in the late 1920s with experimental SW transmissions, and from the early 1930s was a major player with a regular schedule, the BBC-WS has acquired the skills necessary to operate a propaganda service. It has honed and refined those skills so that it is the envy of many of the world's major international broadcasters. Because the BBC-WS claims to be an independent broadcaster and not subject to any form of government control, it is able to broadcast types of programmes to foster such beliefs. Trust is the first hurdle for any propaganda broadcaster to overcome, and to achieve this the broadcaster has to be credible at all times. Credibility and candour will promote trust in the mind of a listener. For example, the BBC-WS has been known to criticise the government of the day and members of the British government. Such actions effectively support the BBC-WS claim of independence.

But whilst the BBC-WS is able to behave in such a manner in its overseas broadcasts, such a technique would be less effective coming from an international broadcasting organisation which overtly proclaims to be the voice of the nation. A further reason why the BBC-WS is so trusted may have something to do with the fact that Britain is not a superpower and no longer commands a great empire; by itself Britain does not pose a major threat to other countries.

In both of these respects Voice of America (VOA) suffers from two disadvantages. It is, and proclaims itself to be, the voice of that nation and, because the US is a superpower, many countries treat anything said by VOA with a certain degree of caution. One thing quickly learned by a writer or journalist in international broadcasting is that inside the walls of the BBC's Bush House, the staff exhibit unease at the BBC being described as a state broadcaster and even greater unease (as noted) when the word propaganda is mentioned. Yet the definition of the word involves projecting a faith, a belief or a viewpoint. The World Service is mandated by the terms of its royal charter to present views on world affairs taking into account British government policy. It is therefore no different to those international broadcasters who are quite happy to be described as propaganda broadcasters.

Being a major propaganda broadcaster with a high profile in global politics requires a knowledge of what other broadcasters in the world are saying, and what is going on in the regions of the world where there is conflict. The BBC-WS operates the best-known, and almost certainly the most efficient, SW monitoring service in the world. On a technical level it monitors almost everything that is broadcast on SW and also MW from almost every region of the world. From this monitoring activity, intelligence digests are prepared and issued on a regular basis to the British government. Of greater significance may be the fact that it also advises the government on the current state of jamming of BBC-WS broadcasts and how it can be countered.

The World Service, as part of the British Broadcasting Corporation, is governed by a Chairman and a board of governors who are appointed by the state and answerable to parliament. The World Service differs from the national broadcasting part of the BBC in that it is funded by the state through a three-year budget administered by the Foreign and Common-wealth Office. Its funding for the years 1991 to 1994 is shown in Table 6.1.

Table 6.1 *Annual budget for the BBC-WS*

Year	£ million
1991–1992	159.6
1992–1993	166.3
1993–1994	175.8

It should be noted that these figures do not include a capital budget for the building of new SW and MW transmitter stations and other capital projects. This is a separate item.

A high-profile international broadcaster

No other international broadcaster puts in as much effort as the BBC to reassure its public that it is a broadcaster of truth, with slogans such as:

> Free and untainted information is a basic human right. Not everyone has it but everyone wants it. It cannot by itself create a just world, but a just world cannot exist without it.

Yet truth is an elusive commodity. Never pure, rarely simple and dependent upon many different factors, not the least of which is the *source* of the truth. Nevertheless the credibility of the BBC-WS has never suffered a serious fracture. In fact it is a rolling success story. Its audited listening base shows listening audiences which have crept up from 120 million during the cold war to an all time high of 143 million in 1998 (shown for 1996 in Table 6.2).

Table 6.2 *BBC-WS listening figures (direct broadcasting, 1996)*

Region	No. of listeners
Americas: Canada, USA, Central and South America	8 million
Africa and the Middle East	42.5 million
Europe including Central and Eastern Europe	12.5 million
Former Soviet Union and S E Asia	7.5 million
South Asia	52.5 million
Asia Pacific Rim countries	7.5 million

These figures exclude certain countries in SE Asia where audited data are not available. According to the BBC these countries are Afghanistan, Myanmar (though the BBC still refer to this country as Burma), Cambodia, China, Cuba, Iran and Somalia. Some of these countries are, according to the US President's Task Force Report, 'The last bastions of communism in the world'.

One which certainly fits that title is the People's Republic of China. It was because of this that the BBC invested in a new state-of-the-art SW station in Hong Kong which came on air on 27th September 1987, to reach out to Central and Northern China with the identity of BBC East Asia Relay, giving no clue as to its exact location on the shoreline of Hong Kong. In June 1997 Hong Kong reverted to Chinese control, which poses several questions: did the SW facility justify its short existence, and what happened to the transmitter station on the shoreline of Tsang Tsui in Hong Kong after July 1997? When these questions were directed to the head of broadcaster coverage in the BBC in 1994, the answer to the first was 'yes' and to the second was 'we don't know'. Subsequently, in 1996 the BBC-WS confirmed that the SW station has been closed down and the transmitters removed.

That the FCO in conjunction with the BBC-WS deemed it to be a sound expenditure of taxpayers' money to invest in a high power transmitter facility in Hong Kong with the knowledge at that time that the facility had a nine-year life, serves to underline the importance attached to international propaganda broadcasting, and that when global politics is the issue the cost to the taxpayer is a secondary matter. There can be little doubt that the SW station at Tsang Tsui was intended to play a pivotal role in broadcasting to the people of China, as it probably did in the Tiananmen Square student uprisings. These student riots failed to ignite a general uprising throughout China, but the demonstrations just might have succeeded. In the event the Chinese government blamed Western influences and has jammed BBC broadcasts to China ever since.

That the BBC-WS is an international broadcaster with a high profile in world events and global politics becomes evident from books written by BBC-WS staff [3, 4], with confrontation being present when broadcasting to certain countries. Certainly the World Service seems to attract more jamming than many other European international broadcasters, including those such as Deutsche Welle, perhaps as a result of confrontational broadcasting. Nevertheless, it is fact that the BBC-WS commands the greatest listening audience in the international broadcast bands and it commands respect from adversary and friend alike [2].

The BBC-WS has been in existence longer than most other international broadcasters, and long enough to develop and hone its techniques. As long ago as the mid 1930s Moscow Radio described it as a 'subtle and jesuitically refined tool for influencing minds'. Compliments do not come much higher than that. More recently the BBC-WS has been co-operating with Russia in many ways, relaying its programmes over local radio stations and assisting with its 'English by radio' learning programmes. It also leases air-time on some of Russia's powerful SW facilities in Siberia and elsewhere for the purpose of targeting certain parts of Asia. That the BBC-WS and the FCO still regard Russia as a major world power is evident in the World Service's expenditure on Russian language programming, which has the second highest budget allocation.

Continuing development

Since the ending of the cold war the BBC World Service has toned down its inflammatory style of broadcasts, which it used to great effect during the cold war. Inflammatory is a word with which the World Service would certainly not agree, but the leaders in the Kremlin would have perceived the broadcasts in this way which caused them to make crippling expenditure on the construction of powerful sky wave and local ground wave jammer installations.

In 1994 the BBC carried out a major restructuring programme. A new organisation, BBC Enterprises, was set up to co-ordinate and manage three departments, BBC World Service, BBC International TV, and BBC Publishing. The World Service became the main bedrock of the new organisation, though distinct from the other two departments in its funding, which continues to come from three-year FCO grants.

Though the popularity and respect which the BBC-WS commands throughout the world is due mainly to its claim to be a non-partisan organisation and free of government control, its quality and style of programming plays a vital role in its popularity – to an extent where even if it were a weak signal in the broadcast bands listeners would probably remain loyal. In fact the BBC-WS is one of the strongest signals in most parts of the world. This is due to its HF network of relay stations augmented by a similar network of high power MW transmitter sites. Supplementing these two networks is a collection of leased transmitter facilities.

Table 6.3 is reproduced from the FCO Capital Review, September 1994, and lists BBC-owned transmitter sites in the UK and abroad. Figure 6.1 shows SW sites and the regions served by these facilities. Figure 6.2 shows leased sites and exchange relay stations such as Delano, Bethany and Sackville in North America. Figure 6.3 shows MW sites owned or leased by the BBC.

It is evident from these three maps that Russian-provided SW and MW facilities, at places like Ekaterinburg, Moscow, Kiev, Tashkent, Irkutsk and Chita, have and continue to contribute significantly to the BBC-WS coverage in some important and politically sensitive parts of Asia and the Middle East.

Table 6.3 *BBC-owned transmitter sites, UK and abroad*

Station	Number and type of transmitters in 1981	at present	First year of operation (for present transmitters)
Antigua	2 × 250 kW SW	2 × 250 kW SW	1976
Ascension Island	4 × 250 kW SW	4 × 250 kW SW	1966
		2 × 250 kW SW#	1963
Crowborough	2 × 100 kW SW	station closed	
	2 × 600 kW MW	in 1986	
Cyprus Ziggi		6 × 250 kW SW	1982
Ziggi	4 × 100 kW SW*		1963
Ziggi		4 × 300 kW SW**	1988
Ladies Mile	2 × 500 kW MW	2 × 500 kW MW	1978
Ziggi	2 × 100 kW MW	1 × 200 kW MW	1958 (Paired)
	2 × 20 kW SW		
	2 × 7.5 kW SW		
Daventry	3 × 100 kW SW	station	
	5 × 100 kW SW	closed	
	4 × 250 kW SW	1992	
Hong Kong		2 × 250 kW SW	1987
Lesotho	1 × 100 kW SW	1 × 100 kW SW	1981
		1 × 100 kW SW	1987
		1 × 100 kW MW	1990
Masirah	4 × 100 kW SW	4 × 100 kW SW	1977
	2 × 750 kW MW#	2 × 750 kW MW#	1961
Orfordness	1 × 50 kW MW		
	1 × 250 kW MW	1 × 250 kW MW	1977
	1 × 500 kW MW	1 × 500 kW MW	1977 (Paired)
		1 × 600 kW MW	1982
Rampisham		4 × 500 kW SW	1983
		4 × 500 kW SW	1985
		2 × 500 kW SW	1991
	2 × 100 kW SW		
	4 × 250 kW SW		
Seychelles		2 × 250 kW SW	1988
Singapore	4 × 100 kW SW#	4 × 100 kW SW#	1969
	4 × 250 kW SW#	4 × 250 kW SW#	1969
		1 × 250 kW SW#	1963
Skelton 'A'	6 × 250 kW SW	6 × 250 kW SW	1968
Skelton 'A'		5 × 250 kW SW#	1964
Skelton 'C'		4 × 300 kW SW	1991
Skelton 'C'		2 × 300 kW SW	1987
Skelton 'B'	13 × 100 kW SW	station closed in 1990	
Woofferton		4 × 250 kW SW***	1963
Totals	75	75 (excluding 2 × Txs owned by RCI)	

Notes:

\# Transmitters moved to present location from another site.
*   Currently being replaced by 4 × 300 kW ex-Daventry.
** Ex-Daventry transmitters not yet in service.
*** ex-VOA transmitters.
Source: FCO Capital Review, September 1994.

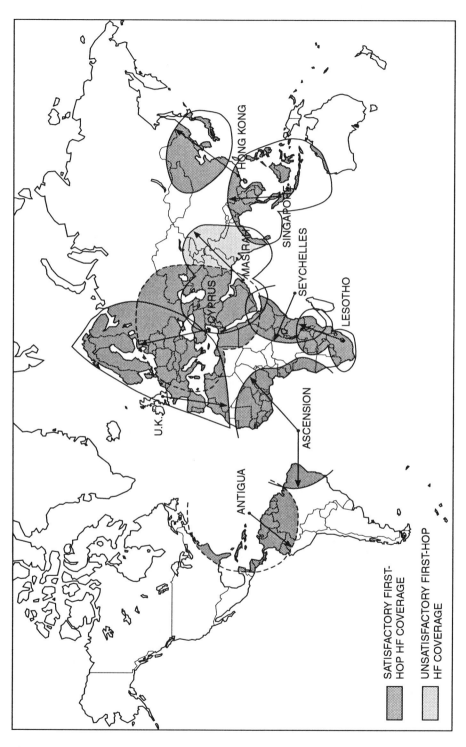

Figure 6.1 *BBC World Service HF transmitter facilities and coverage for transmitters owned by the BBC*

HONG KONG

SINGAPORE

MASIRAH

CYPRUS

SEYCHELLES

LESOTHO

U.K.

ASCENSION

ANTIGUA

SATISFACTORY FIRST-HOP HF COVERAGE

UNSATISFACTORY FIRST-HOP HF COVERAGE

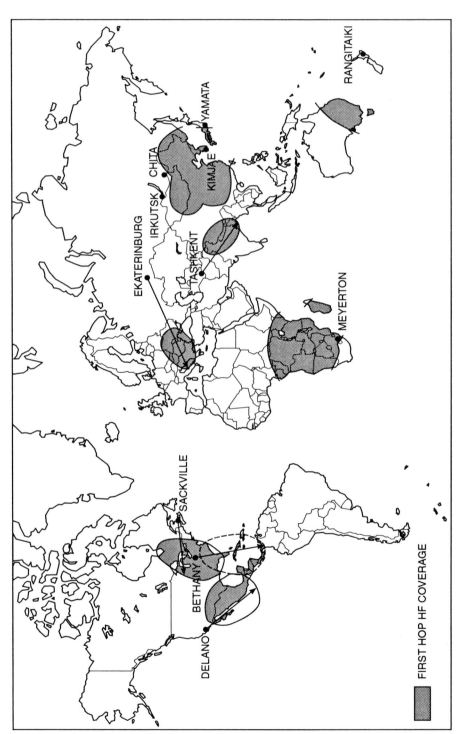

Figure 6.2 *BBC World Service HF transmitter facilities and coverage for transmitters hired by the BBC*

FIRST HOP HF COVERAGE

DELANO

BETHANY

SACKVILLE

EKATERINBURG

IRKUTSK

CHITA

TASHKENT

KIMJAE

YAMATA

RANGITAIKI

MEYERTON

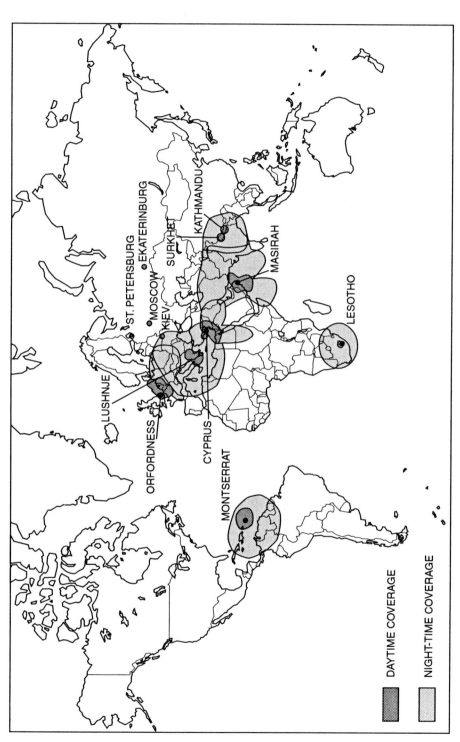

Figure 6.3 *BBC World Service MW facilities, showing daytime and night-time coverage*

DAYTIME COVERAGE

NIGHT-TIME COVERAGE

LUSHNJE

ORFORDNESS

CYPRUS

MONTSERRAT

ST. PETERSBURG

EKATERINBURG

SURKHET

KATHMANDU

MOSCOW

KIEV

MASIRAH

LESOTHO

British Government audit of the World Service

The findings of the Comptroller and Auditor General of the National Audit Office (NAO) into the workings of the BBC-WS were officially published in November 1995. (National Audit Office. HMSO HC 111, 15 November 1995, London.) The service was last examined by the Committee of Public Accounts in July 1992. The 1995 report runs to 42 pages of text, with many performance tables, a summary and conclusions. The report deals with such matters as studios, programme costs, overheads, shared services, efficiency reviews, income, cash-flow, audibility, organisational developments since 1992, efficiency and capital programme impact. Information in the report gives the reader facts about transmission costs per kW hour, costs per broadcast, total programme hours, income, numbers of staff, studio hours per transmission hour, studio utilisation, performance in capital projects and so on.

For figures and statistics this is the document to read. On the other hand, should a reader be seeking an accurate, clear picture of why the World Service exists, what it does, how it does it and, more important, why it does it, then this report will leave them no wiser. Most experts would agree that the means of delivery – the transmitter facilities scattered around the world, and why they are in the places they are – is the most important part of the overall structure of a SW broadcaster, yet nothing on this aspect is covered in the Audit.

The report did not define international broadcasting, the number of international broadcasters nor their roles in world politics. It also failed to describe why SW in the HF spectrum is the chosen medium for such broadcasting, and though indirect broadcasting and direct broadcasting operations were briefly mentioned, the differences between the two were not described. All of these omissions could be justified if the report was published for the benefit of those working in this medium but this was not so. The report was published not only for the BBC but for public consumption.

Yet the public (certainly in the UK) has little or any knowledge of SW broadcasting and generally believes that it is old-fashioned, or that it is something practised by radio amateurs and hobbyists. Such ignorance is not exclusive to Britain; the same perceptions are held in the pluralistic societies of the US and in Western Europe. They are in the main completely unaware that there are large regions of the world, especially in the Middle East and Asia, where SW listening remains a staple, because it is used for national broadcasting in many countries. The value of the Audit Report both to the general public and Members of Parliament would have been higher had such information been included.

Nevertheless, the report does contain information of high value to many in the international broadcasting sector. There is, for example, a table of relative running costs for SW relay stations. Rampisham, the BBC-WS SW station in England, was operated at an annual cost of £3 million.

Table 6.4 *BBC World Service statistics*

Broadcasting hours	896 per week direct broadcasting and 563 hours indirect broadcasting (April 1995 figures)
Numbers of transmitters	78 transmitters on 10 BBC-owned sites plus another 80 transmitters and relay stations in more than 30 countries
Global listening audience	133 million +
State funding from FCO	£178.8 million for 1995–1996 and 1996–1997
Transmitter kilowatt capacity	20,900 million kW (an increase of 73 per cent over 10 years)
Audibility program	a 10-year audibility improvement programme was completed in 1991 at a cost of £166 million
Further planned extensions	A new SW station in Thailand is to have four 300 kW SW transmitters. [This came on air to replace the SW station in Hong Kong.] The existing two 750 kW MW transmitters on Masirah island are in need of replacement.
BBC World Service staffing	2183 with plans to reduce to 2100 by 1996

Source: National Audit Office. HMSO HC 111, 15 November 1995, London

Rampisham is equipped with ten 500 kW SW transmitters and by assuming a usability factor of 17 hours per day for nine out of ten transmitters (one on standby) the hourly operating costs per transmitter, exclusive of capital cost, work out to £56 (US $90). This is useful information when assessing the merits of leasing transmitters from another broadcaster.

One other highly informative element of the report is a detailed table of the relative production costs for the language services broadcast by the World Service. French language costs are less than £100 per hour, whereas Albanian programming is reported at £1859 per hour, at eighteen times the cost of French language programmes. This fact alone serves to emphasise that international broadcasting is a political tool. For what other reason could it be justified for the BBC-WS to go to such extreme costs to serve what is generally accepted to be the poorest state in the whole of Europe?

The Audit Report goes on to say that the World Service is considered to be the best known and the most reliable of all international broadcasters, with an estimated regular audience of 133 million, at the time of the report. Though few would disagree with such statements the report should have attached credit to the English language which commands a greater world market than any of the other widely spoken languages.

Some would argue that the value of the Audit Report would have been greatly enhanced if, instead of going into fine detail on the BBC World Service operations, it had produced a table of useful parameters which compared the World Service against other Western broadcasters, such as Radio France International, Swiss Radio International, Deutsche Welle and Voice of America, showing (perhaps) annual budgets against measured world audiences.

Chapter 7
Deutsche Welle:
the Voice of Germany

As befits the superpower of Europe, Germany has more public service broadcasters (PSB) than any other country in Europe. But even before the former East Germany merged with the Federal Republic, West Germany was already the biggest broadcaster in Europe, whilst the city of Berlin itself, due to geopolitical factors, became the cockpit for East–West relations and home to foreign radio stations operated by American, British, French and Russian authorities – some of which are still in place.

ARD (Arbeitsgemeinschaft der offentlich-rechtilchen Rundfunk – anstalen der Bundesrepublik Deutschland) is the umbrella organisation that looks after the interests of these PSBs and includes amongst its members Bayerischer Rundfunk (BR), Deutschlandfunk (DLF), Hessinger Rundfunk (HR), Mitteldeutscher Rundfunk (MDR), Norddeutscher Rundfunk (NDR), Ostdeutscher Rundfunk Brandenburg (ORB), Radio Bremen (RB), Saarlandischer Rundfunk (SR), Suddeutscher Rundfunk (SDR), Sudwestfunk (SWF), Westerdeutscher Rundfunk (WDF), Deutschlandfunk (DLF) and Deutsche Welle (DW). These public service broadcasters radiate on all the radio broadcast bands and some operate as many as fifty radio stations.

Unique in ARD is Deutsche Welle, the German international broadcasting service. As is the case with practically all the international broadcasters Deutsche Welle is not financed by licence fees but through federal funds, that is, taxpayers' money. Under the terms of its charter it has the task of conveying in its programmes an accurate and comprehensive picture of the political, cultural and economic life in Germany to its audiences abroad. But as might be expected, the more Europe becomes unified the more the European component becomes a significant part of its programme. Beyond the terms of its charter the Deutsche Welle service is independent and free from government control. Nevertheless it has to ride that fine line between complete independence from the federal government of the day and broadcasting programmes that are in the interests of the state as a whole.

Figure 7.1 *High power UHF-TV transmitter station in the Austrian Alps*

When analysed in 1996 Deutsche Welle transmitted programmes in German and 39 other languages to every part of the world, with a weekly total of 660 programme hours per week. According to the figures then available from the International Broadcast Audience Research Unit, Deutsche Welle ranked fifth in the world, behind the USA (VOA, RFE and RL) with

Figure 7.2 *600 kW MW transmitter of Radio Bavaria, circa 1970*

2008 hours, China with 1673, Russia with 1332 and the United Kingdom with 878. In the decades between 1970 and 1990, however, Deutsche Welle ranked fourth, broadcasting more programme hours than the BBC – with the end of the cold war Deutsche Welle cut back its programme hours by nearly 25 per cent.

History of German SW broadcasting

Germany was one of the first nations in Europe to develop a strong broadcasting capability in the then developing technology. The mid 1920s saw the inauguration of its first long distance service. This was from Nauen to Argentina on a frequency of 4 MHz and the transmitter output power used on this service – a mere 0.8 kW – conveys proof of how long range broadcasting could be achieved with flea power when compared to today's 500 and 1000 kW SW transmitters. The reason for this is that the ether was uncluttered, and man-made atmospheric and electromagnetic noise was almost non-existent since many areas were largely without electricity.

Table 7.1 *SW and MW transmitter sites for Deutsche Welle*

Transmitting stations within Germany

Julich	11 SW transmitters – each 100 kW	
Wertachtal	9 SW transmitters – each 500 kW	
,,	4 SW transmitters – each 500 kW	(leased to VOA)
,,	2 SW transmitters – each 500 kW	(spares)
Nauen	3 SW transmitters – each 500 kW	
,,	1 SW transmitter – 100 kW	

Transmitting stations overseas

Kigali, Rwanda	4 SW transmitters – each 250 kW
Cyclops, Malta	3 SW transmitters – each 250 kW
,,	1 MW transmitter – 600 kW
Sines, Portugal	2 SW transmitters – each 250 kW
Sri Lanka	3 SW transmitters – each 250 kW
,,	1 MW transmitter – 600 kW
Antigua, Caribbean	2 SW transmitters – each 250 kW

Overseas stations on lease or on an exchange basis

Sackville, Canada	2 SW transmitters – each 250 kW
Brasilia, Brazil	2 SW transmitters – each 250 kW
Novosbirsk, Russia	2 SW transmitters – each 1000 kW
,,	2 SW transmitters – each 100 kW
Samara, Russia	1 SW transmitter – 250 kW
,,	1 SW transmitter – 200 kW
Irkutsk, Russia	1 MW transmitter – 250 kW
Moscow, Russia	1 MW transmitter – 10 kW
St Petersburg, Russia	1 MW transmitter – 10 kW
Sofia, Bulgaria	1 VHF-FM transmitter – 1 kW
Montserrat	1 MW transmitter – 135 kW

1926 saw the start of SW broadcasting from Nauen to the USA, Central and South America and the Far East. Nauen was soon followed by a second state-of-the-art SW station, Zeesen, which became the flagship transmitter. Between the late 1930s and 1945 Germany was the most powerful and technologically advanced broadcaster, and from 1940 onwards the Third Reich exercised effective control of a radio broadcasting network that stretched from the North Cape as far south as North Africa, and from France to Poland and the Balkans.

Defeat in the war saw a temporary end to German SW broadcasting. Regular SW broadcasting from the Federal Republic started up again on the 3rd May 1953. By 1955 its programme hours were a modest 105 – less than a quarter of that from the BBC at that time, but by 1965 Deutsche Welle had not only overhauled the BBC-WS but was once again the most powerful broadcaster in Europe, reaching a peak with 848 programme hours by the end of the cold war. This compares with the figure of 408 programme hours from Radio Berlin International, its East German competitor.

Transmitter network

In its early years, from 1956, Deutsche Welle had one high power SW station, Julich. It had eleven 100 kW Telefunken transmitters and the station remains in service in the 1990s. In 1962 the Federal Government authorised the German Federal Post (DBP) to commence planning for a new state-of-the-art SW transmitting station. The result was Wertachtal, which went on air in 1972 with four 500 kW SW transmitters, but was destined to become – no doubt partly fuelled by the cold war – the most powerful SW transmitting station in the Western hemisphere by 1989 with fifteen 500 kW transmitters, four of which later went on to relay Voice of America broadcasts. As a study in technical excellence, even in the late 1990s, this powerful SW station has very few equals.

Deutsche Welle now radiates its programmes over 39 SW and two MW transmitters. Three of these stations are in Germany and it has five relay stations overseas in Africa, Malta, Portugal, Sri Lanka and in the Caribbean. In addition it leases transmitters on Russian soil and has exchange relay agreements with Canada, Brazil and Montserrat.

From its SW stations in the Federal Republic Deutsche Welle radiates programmes to many parts of the world. The Nauen site in the former GDR was inherited from the East Berlin broadcaster, Radio Berlin International. Its three 500 kW and one 100 kW transmitters of Russian manufacture have been replaced with four of the latest 500 kW SW transmitters from Telefunken and it now also has four rotatable ALLISS antennas from Thomcast. With Nauen on-air Deutsche Welle had a total of nineteen

Figure 7.3   *Wertachtal SW station, Bavaria, Germany*

500 kW transmitters, including the four leased to VOA from the Wertachtal station, plus the three spares.

Coverage of the African continent is achieved by the Kigali station in Rwanda, whilst the three SW transmitters and one MW transmitter on Malta serve the Mediterranean countries and Near East and give partial coverage of North and South America, Southern Asia and Far East. The transmitting station in the Caribbean is a joint-venture with the BBC and serves North and South America. In 1989 another gap-filling operation took place when a station was constructed in Sri Lanka to cover the Asian regions. Additionally Deutsche Welle has a relay-exchange agreement with CBC Canada which permits Deutsche Welle air-time on two 250 kW SW transmitters.

As a result of negotiations with Russian authorities following the end of the cold war, Deutsche Welle has since the autumn of 1991 been using some very high power SW transmitters on Russian soil to broadcast into China and regions of Asia. These sites are located at Novosibirsk, Samara and Irkutsk. Two of the transmitters are of 1000 kW carrier power, confirming that very high power SW technology in the Soviet Union during the cold war was as advanced as in the West, and probably even more so when it came to building very high power SW transmitters.

Though Western broadcasters like Deutsche Welle, the BBC-WS and others collaborated with their broadcasting to the Soviets during the cold war years, international broadcasting in the HF spectrum is nevertheless a highly competitive business in the sense that many compete for the same listening audiences. The post-cold war years have seen some far-reaching changes in programming as Western broadcasters have re-adjusted to changed circumstances following the dismantling of communism. SW broadcasters are innovating, for example, by offering other countries the opportunity of re-broadcasting their programmes over local FM networks or by allowing direct delivery of their programmes by satellite to hotels. This is called indirect broadcasting. Such arrangements make it possible for broadcasters like Deutsche Welle to increase their audience listening figures without the expense of constructing more SW transmitter sites. Another way of increasing audiences is by offering selected partners transcription programmes in the form of tapes or audio cassettes.

Deutsche Welle has been particularly successful in both mediums. In 1993 it supplied some 170,000 programme copies to more than 1200 partner stations all over the world. In addition almost 1000 educational institutions in Eastern Europe are supplied with German-language information programmes for use as teaching aids. This is an excellent and cost-effective method of projecting a nation's culture and way of life to foreign audiences whilst also prompting foreign audiences to tune directly to Deutsche Welle broadcasts via the short and medium waves.

Conscious of the fact that foreign audiences for SW listening are to be found mainly in the developing countries of Africa and Asia, and that SW is a shrinking market in the developed countries of the West, Deutsche Welle in April 1992 started up a satellite service to selected regions of the world, broadcasting current affairs programmes in three languages, German, English and Spanish. The programmes were transmitted 14 hours daily (16.00 to 06.00 UTC) over various satellites to different parts of the world. (See Table 7.2.)

Table 7.2 *Deutsche Welle satellite usage*

Satellite	Region
EUTELSAT 11-F	Europe, Middle East, North Africa (plus radio via audio sub-carrier, German programme in stereo)
INTELSAT-K	North, Central and South America (plus radio via audio sub-carrier, German programme in stereo)
SATCOM C-4	North America, Caribbean (plus radio via audio sub-carrier, German programme in stereo)
SPACENET 11	Transmits $2\frac{1}{2}$ hours to North America and Caribbean
INTELSAT 505 } INTELSAT 601 }	4 hours to Africa and parts of Asia
INTELSAT 508	$2\frac{1}{2}$ hours to East Asia, Australia and New Zealand

Deutche Welle (in common with practically all other international broadcasters) is chiefly financed by the Federal Government. Back in 1964 it employed over 2000 permanent staff in its editorial, engineering and administrative sections. In 1993 its budget was DM 605 million and this increased to DM 658 million in 1994. This figure is one of the highest budgets ever for an international broadcaster.

Wertachtal SW transmitter station

Wertachtal is the current flagship and hub of Deutsche Welle's SW system. Equipped with fifteen high power SW transmitters of the 500 kW type, twelve of which are in operational service, with three on standby or as spares, this SW transmitting station is one of the largest in the world. Indeed it may be the largest in the world because it is doubtful that Balad in Iraq, with sixteen 500 kW transmitters, has been 100 per cent operational since the Gulf War. But even with this capability Deutsche Welle sees Wertachtal as eventually in need of expansion to accommodate something like twenty transmitters.

Wertachtal first came on air in time for the 1972 Olympic games, with an initial fitment of four transmitters. Planning for the new station started ten years earlier in 1962. Before 1972 Deutsche Welle's international SW service had been carried by its SW station in northwest Germany at Julich, with eleven 100 kW SW transmitters. The increasing importance of SW broadcasting as an instrument of foreign policy at that time, augmented by the fact that the Federal Republic was becoming a main player in the cold war, led to Deutsche Welle planning a higher-powered SW site in conjunction with the Deutsche Bundespost Telekom (DBP).

From the outset the station design concept provided for the construction of a SW station with a goal of twelve 'superpower' transmitters with carrier output power of 500 kW, fitted with auto-tune facility and able to operate on any frequency within the range from 5.9 to 26.1 MHz. The concept further provided for a total of 76 SW antennas with the capability of being able to select any one of the transmitters to any one of the antennas. This was to be accomplished with a cross-point switching matrix of 912 RF switches, to give a switching capability of 12 transmitters to 76 antennas.

This was certainly the highest-powered SW station envisaged at that time, and called for some stringent requirements relating to the selection of a suitable location. After an exhaustive search for a site DBP was offered a site just south of Augsburg with the right topography, good electrical conductivity and good isolation from urban development.

Wertachtal is very advanced in its antenna system and RF feeder design. Is uses coaxial feeders throughout, from transmitter to antenna. This concept came about because of an environmental factor. Wertachtal is located on the lower slopes of the Bavarian Alps; the region is characterised by temperature extremes, solar hazards and subject to gale force winds, driving rain, very heavy snowfall and severe icing. Faced with such harsh environmental conditions, Wertachtal became the first high power SW station to use a coaxial cable transmission line as an alternative to tradi- tional open wire feeder lines which would be subject to potential flashovers. The entire transmission line system is at 50 ohms impedance from trans- mitter to antenna, with no need for impedance transformation baluns – as is the case with open wire balanced feeders. When put into practice the con- cept was found to have many advantages. Apart from the more obvious, such as avoidance of mismatches, reduction of near-field effects and unwanted radiation, the system has much greater reliability; not a single major fault has occurred from the time the station was completed.

Antenna layout

The layout of the antenna system takes the form of three bent radial legs spaced at 120 degrees. Stacked dipole arrays are supported by 25 steel lattice towers. Each pair of towers supports two curtains back to back with a wire screen reflector interposed between the two curtains giving each pair of masts a reverse direction of fire capability. Additionally each curtain array has a continuous slew capability of ±30 degrees from the straight ahead angle of azimuth fire. Commands for the control of the slewing switches are issued by a process computer; the line lengths of the phase elements are designed to be infinitely variable, but to ease system design they have five positions within the total range of 30 degrees, each selection is achieved by a single command, thus each curtain requires five different commands.

The entire transmitter station is computer-controlled. The heart of this is a screened room with three process computers, one of which is the back- up system, continually updated with the current operating data so in the event of a failure in the operational computer, the back-up system will take over. The station is menu-driven to a pre-arranged schedule so that at an appointed time the system will select a transmitter and initiate auto-tune to the frequency desired, select the antenna and its direction of slew.

Transmitters are of two types, both of Telefunken manufacture. One thing not seen at Wertachtal is a motley collection of equipment of different manufacture. Earlier transmitted are of type S2500, this model being Tele- funken's first generation of 500 kW transmitters, and the remaining are type S4005, a later development. As the earlier model requires more floor area than the S4005 they are being replaced with newer models from Telefunken, allowing Deutsche Welle to have sufficient space in the transmitter building to accommodate twenty 500 kW transmitters. It is likely that the size of the antenna switching matrix will also be extended.

Though this SW transmitter station was conceived and planned in the 1960s it remains one of the most advanced anywhere in the world and it extended greatly the frontiers of knowledge in SW transmission technology. The high expenditure in devising a transmission system that is RF tight, radiation-proof and hermetically sealed from transmitter output to the point of connection to the antenna, has been more than justified by the extraordinary standard of performance and reliability which Wertachtal has achieved.

The design and construction of a fully automated, multi-frequency, high power, menu-driven SW transmitter station calls for knowledge and skills of a high order in many different disciplines. This was an all-European project in which many experts from Deutsche Bundespost Telekom and other organisations involved. Telefunken supplied and engineered the high power transmitter system and associated control systems, with the Mannheim Division of ABB responsible for the antennas and associated items.

The revival of Nauen: Europe's newest SW station

The world's first superpower transmitting station, created almost 100 years ago, came back on air at Nauen, 40 km outside Berlin, on the 25th April 1997, and to commemorate that historical event, this brand new SW station was chosen as the place from which to broadcast the world's first digital–analogue transmission by SW. Representatives of the State of Brandenburg, Deutsche Telekom and the broadcast industry and media were present to witness this historic event.

The history of Nauen has its origins in the late nineteenth century when two eminent German scientists from the Allgemeine Elektrizitats (AEG), Professor Adolph Slaby and Count Von Arco, were in competition with Marconi in England, and with Ferdinand Brauen of Siemens and Halskie (also of Germany). The heads of the two Berlin-based companies agreed to work together in order to compete more effectively with the English company. Thus, on 27th May 1903, Emil Rathenau and Werner von Siemens founded the jointly-owned company Gesellschaft fur Drahlos Telegraphie, The Wireless Telegraph Company, with the trade-mark Telefunken. First Telefunken was based in the city of Berlin, but shortly afterwards it moved to the small town of Nauen, 40 km to the north-west.

The site selected for experimental broadcasting was a 40,000 square metre area of swampland with the water table just below the surface, thus offering perfect conditions for electrical conductivity of wave propagation. By 1909 Nauen wireless telegraph station was on air with a spark transmitter delivering about 10 kW to the antenna. In 1908 the station set its first record with its signals picked up by a ship near Tenerife, 3600 km away. The second phase in the history of Nauen lasted from 1909 to 1911. By this time Nauen was communicating with Germany's African colonies several thousand kilometres distant using an antenna power of 35 kw from Telefunken's own resonant quenched spark gap LW transmitter.

The third phase of the early history of Nauen dates from 1911 when the station went from being an experimental station to a commercial telegraphy station. The dream of Hans Bredow, the station director, was about to be fulfilled: to make Nauen a superpower transmitter. For this a new antenna 260 m high was erected, and the transmitter output power was increased to 100 kW. This transformation enabled the station to play an increasingly important role in the outbreak of World War I in contacting German ships around the world and so guiding them to safe neutral harbours. As the war progressed the British began to sever all Germany's undersea telegraph cables to the rest of the world, and so Nauen was now the sole means that Germany had for communicating with its colonies in Africa and elsewhere. The output power was further increased to 150 kW. Towards the end of the war the station at Nauen was the subject of an attempt by the USA to bring about the collapse of Germany by subversive propaganda. During the course of the war American telegraph stations had monitored transmissions from Nauen, but had never attempted to communicate until then. The American telegraphy station at New Brunswick started to call Nauen: 'POZ-POZ-POZ de NFF'. Nauen responded in telegraph code 'NFF de POZ, ur sigs FB OM', 'Your signal's fine old man'. Startling as the response was, the message that came back from NFF was even more so because it was nothing less than a message from President Wilson addressed to the German people asking them to overthrow the Kaiser and so end the war. This was the first time in history that the US administration attempted to bring about an uprising by radio.

After the war, Nauen went on to become of greater importance. It developed from being a LW sender to a SW station when the short waves superseded long waves for long range communication. On 1st January 1932 the German Reichpost, acting on the government's authority, assumed control of Nauen, its superpower transmitter and 250 hectares of land. From that point on the station, with its buildings, antennas and transmitters, expanded considerably and by 1939 Nauen was one of the biggest and most powerful communication complexes in the world. It had SW transmitters but also LW transmitters to enable it to send messages to the German U-boat fleet. The station was never damaged by allied bombers but disaster of another sort struck on 24th April 1945, when the town of Nauen was occupied, along with the Nauen complex, by the Russian Red Army. Russian soldiers dismantled the high power transmitters and dynamited the antenna masts, although the main building survived.

Redevelopment for the cold war

For a few years Nauen sank into oblivion, and it took the cold war to revive the fortunes of this historic station; from 1951 to 3rd October 1990 Nauen transmitter station served the German Democratic Republic in a number of ways. From 1951, three years into the cold war, the transmitter station was under the control of the East German post office and was initially used for telephone communications between Oranienburg, Berlin and Falkensee. But with the founding of the Nauen Radio Office on 1st January 1956 the station resumed its position as a superpower station. From 1959 to 1989 21 transmitters with powers up to 100 kW and 45 different antenna systems were installed for worldwide commercial radio communications. Nauen served a number of government departments, including the Central Telegraph Office, the Overseas Department, the Potsdam Geodetic Institute, the Weather Centre, the Marine Navigation Service and, not least, the East German Ministry of Foreign Affairs.

On 15th October 1959 Nauen was considerably extended to enable it to function as an overseas broadcaster, Radio DDR. Under the name Radio Berlin International (RBI), foreign service broadcasting was carried out. At first it had a single 50 kW transmitter, but like all other international broadcasters RBI was an important player in the cold war, and from 1959 further expansion took place. Two East German state-owned companies, VEB Funkwerk Berlin Kopenich and VEB Industrieprojektie, developed a state-of-the-art swivel-rotatable SW antenna at Nauen. Built as a prototype, it served the transmitter station from 1964 until the end of the cold war. Even today, according to Telefunken and Deutsche Telekom, this antenna is unique and remains in operational use.

The design consists of two rigid dipole arrays that can be used to cover broadcast bands from 5.8 to 18.8 MHz. It can be rotated through 360 degrees of azimuth, and tilted from zero to 50 degrees vertically with a positioning accuracy of 0.5 degrees. It has a power handling capacity of 200 kW, a forward gain of 14.1 to 20.0 dB, weighs 285 tons and is capable of fully automatic operation by remote control. It can make all changes in both azimuth and vertical angles in less than five minutes. This antenna, designed as long ago as 1959, is yet another example of how far the Soviets and their allies had progressed in superpower broadcast transmission science.

When the antenna went into service on 2nd February 1964 it marked the establishment of the German Democratic Republic's shortwave centre (KWZ) and it made Nauen and Radio Berlin International the second most powerful foreign service in the Eastern bloc after Moscow.

Between 1971 and 1981 Nauen was further upgraded by the addition of three new 500 kW super transmitters, one purchased from Brown Boveri of Switzerland and two from Russia; this was at the time when few of the major Western broadcasters like VOA and the BBC had super transmitters. In parallel, 23 high gain SW curtains designed for world coverage were erected in the shape of an S-plan to ensure optimum coverage of the important target zones. With this powerful concentration of 500 kW SW transmitters and curtain arrays, RBI played a powerful role in the projection of news to the allied powers and to the soviet countries in the Eastern bloc.

German reunification

On 3rd October 1990, the day that the German Democratic Republic was re-united with the Federal Republic, RBI ceased all operations. The Nauen SW centre, including all transmission facilities, was taken over by Cologne-based Deutsche Welle under a contractual agreement provisionally valid until the year 2016. On 1st January 1991 the Nauen station was taken over by Deutsche Bundespost Telekom (the German Federal telecommunications service).

Thus Nauen SW station became the third SW transmitting station of DW and Deutsche Telekom, joining Wertachtal and Julich. At the same time it was decided to upgrade the station with completely new transmitters and new antennas. A contract was awarded in November 1994 to Telefunken Sendertechnik and Thomcast GmbH, with Telefunken as the main contractor. The ability of these two companies to work together in complete harmony had been well demonstrated in the past, with joint projects at Wertachtal and Julich. Under the terms of the joint contract, project leadership was with Telefunken and involved the supply of four 500 kW SW transmitters, type S4105, plus four rotatable SW antennas manufactured by Thomcast GmbH, including a control room for remote operation and supervision of the complete station. The value of the contract was DM 52 million and it was completed ahead of the contract commitment of 28 months. The design philosophy followed that of the ALLISS system, adopted by Radio France International. This is a de-centralised solution, with one 500 kW SW transmitter coupled to its own dedicated rotatable curtain array. The foundations of the 200 ton rotatable curtain array form an open room into which the 500 kW transmitter is installed. The concept possesses much merit in that it eliminates the necessity of an antenna matrix and long coaxial transmission feeders and has since been adopted by three of the major international broadcasters.

Deutsche Welle was the first broadcaster to take delivery of Telefunken's S4105 500 kW SW transmitter. This transmitter is unique for its extraordinary high degree of linearity in audio response which is as high as 30 kHz. Whilst this level of performance may not be essential today, it will be necessary in 20 years' time, or even less depending upon when digital broadcasting is introduced to the SW bands in the HF spectrum. Other outstanding features of the S4105 include a high overall efficiency of 75 per cent and fast tuning to any new frequency in the broadcast bands (within 15 seconds). The S4105 benefits from having new technological features, but at the same time incorporates well-proven techniques used in two earlier generations of 500 kW SW Telefunken transmitters. Much of the transmitter design incorporates solid state with the exception of the final RF amplifier stage, which uses the TH 576 tube from Thomson Tubes Electroniques. This, together with Telefunken's solid state multi-phase PDM modulator accounts for the outstanding overall electrical efficiency. The PDM modulator is characterised by a spectral purity which exceeds all international standards. The S4105 transmitter is able to take SW broadcasting at Nauen well into the 21st century.

Note

The author thanks Johaan Strohmayer of the Deutsche Bundespost Telekom division at Augsburg, Germany, and Dipl.Ing. Jürgen Graaff, managing director of Telefunken, Berlin for assistance with this material on Nauen.

Chapter 8
Radio France International

In common with much of Europe, France has an efficient state-funded broadcasting service. In mainland France radio and television broadcasting comes under two separate authorities; Radio France and Radio France International (RFI). Both have their headquarters and main studios on Avenue President Kennedy in Paris. The former is responsible for regional and national broadcasting, leaving RFI as the programming authority for international broadcasting. RFI ranks with VOA, the BBC-WS and Deutsche Welle as one of the world's major influential broadcasters on the short waves.

Underpinning these two broadcasters is the state-owned transmission authority TéléDiffusion de France (TDF), which is responsible for all engineering activities including the maintenance and operation of all radio and television transmitting stations and links within France plus a few overseas. This arrangement of having a separate authority to run the transmission networks does seem to offer many advantages, and is one that some other countries could adopt and benefit from. Where the same authority is responsible for both the programming and transmission engineering, given the importance of the former, the engineering side can become a secondary consideration.

Of course, the French arrangement can only operate to best advantage when both parties work in close harmony to achieve the end objectives. TDF could be a role model to other countries because it successfully integrates the programme authority with the related objective of TDF, which is the pursuit of technical excellence and to deliver transmissions exactly where intended with maximum audibility and quality of signal, whilst developing a transmission infrastructure that will take France to the year 2020.

TDF has its headquarters in Montrouge, near Paris, with five regional centres: east in Nancy, centre east in Lyon, south-east in Marseille, south-west in Toulouse and west in Cesson Sevigne. The vast transmission network it controls makes TDF the biggest transmission authority in Europe.

Figure 8.1 *Advertisement in WRTH for Radio France International*

France is a nation strongly concerned with science and technology, (indeed, it gave the word 'Ingénieur' [engineer], to the world) and that devotion to engineering is ever-present in TDF.

Like other international broadcasters RFI may be classified as a propaganda broadcaster because part of its function is to reflect and articulate foreign policy of the French government of the day. But RFI differs from other major broadcasters such as VOA or the BBC-WS because it is essentially culture-driven. It took no part in the cold war, nor did it suffer from Soviet jamming operations at the height of that war. Unlike the UK, France does not permit the stationing of foreign troops or propaganda broadcasting stations on French soil.

RFI's policy of treading the fine line of political neutrality, and concentrating on being a culture-driven broadcaster, is believed to stem from the wisdom and foresight of former President de Gaulle himself. He realised the potential that it offered, and instead of broadcasting in huge numbers of different languages RFI broadcasts mainly in French and fourteen others. Much of the programme output is directed to France's former colonies along with other French-speaking countries in the Caribbean and parts of the Pacific. Tangible proof of the popularity of RFI broadcasts and what it achieves for France is in evidence in many parts of the world. In exports, for instance, specifically in the business of broadcast equipment, some 31 countries in the continent of Africa have purchased French-manufactured transmitters and transmission systems from the Thomson group. [2]

Whilst RFI is a culture-driven broadcaster on the international SW bands, it is firmly dedicated to the use of the latest and most efficient radio broadcasting technology. RFI's President, Andrie Larquie, aims to make it the most technologically advanced broadcaster by the year 2000 and, in conjunction with the former chairman of TéléDiffusion de France (TDF) Xavier Gouyou Beauchamps, he jointly signed a contract with Thomson-CSF for the construction of the initial 15 ALLISS systems. That contract, signed in December 1991, followed an earlier contract in January 1991 for a similar quantity of Thomson-CSF's latest 500 kW SW transmitters, to be installed in the ALLISS systems.

The ALLISS concept

This design is based on a concept which reversed all previously held notions on the design and planning of SW complexes. The classic design of a SW station developed over several decades and was based around a central transmitting hall, with another large building to house the giant cross-point switching matrix. The third main element of the traditional design was the transmission feeder system which can sometimes run to 100 km of feeders.

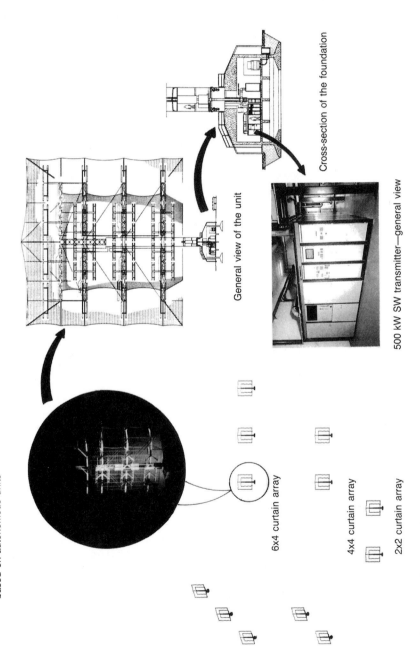

Examples of configurations of a SW centre based on autonomous units

6x4 curtain array

4x4 curtain array

2x2 curtain array

General view of the unit

Cross-section of the foundation

500 kW SW transmitter—general view

Figure 8.2 *Example configuration of an ALLISS SW centre*

Figure 8.3 *First ALLISS project under construction, 1993*

Figure 8.4 *First ALLISS project nearing completion, 1993 (tall structure to the left of the antenna is a crane to assist with rigging operations)*

A disadvantage of this classical design is that some of these feeders can add as much as 2 dB loss in transmitter output power. However, the major disadvantage of this once classical design of a SW complex is the fact that the broadcaster has to estimate what the operational targeting requirement will be in the future. This is because the fixed directive curtain array has a limited slew ability. In the days of the cold war the main target for VOA, the BBC and others was the USSR and Eastern Europe. When the cold war ended, the collapse was sudden and unexpected. The Gulf War brought an urgent need for international broadcasters to make some rapid changes in the direction of their broadcasts, which in practice proved to be difficult or even impossible. Had the concept of ALLISS been in general use from 1989, broadcasters such as VOA and the BBC-Ws would have been able to re-target their transmitters within a maximum time of three minutes. Finally, ALLISS also has the enormous advantage that a SW complex made up of a number of spatially-separated ALLISS silos is a much more difficult target to disable by air-strikes in time of war.

From a production cost standpoint the ALLISS concept is relatively 'modular', based on standard parts, which means considerable cost savings for a complex of say 12 ALLISS stations over a traditionally designed SW station. TDF engineers are convinced that the method will have consider-able appeal to other international broadcasters and the financial investment by RFI for its ALLISS stations underscores the views of TDF.

The hub of the SW network that radiates the programmes of RFI to the world is located in central France at two main transmission sites a few kilometres apart at Issoudun and Allouis. These are operated and main-tained by TDF engineering staff. The sites have a total of twelve 100 kW, one 4 kW and eight 500 kW transmitters. All are SW and were installed in the 1970s. Antennas are of the directive curtain array type, supplemented with two omni-directional antennas. The latest strategy of RFI is to have each ALLISS, as it is completed, take over the services of the existing trans-mitters. The original quantity of fifteen ALLISS systems has now been reduced to twelve. The fact that each ALLISS can be made operational as and when it is completed underlines one of the most important advantages of ALLISS over the classical design approach which takes many years from planning to completion.

RFI's worldwide listening audience

When the SW stations at Issoudun and Allouis were built in the 1970s they were amongst the most advanced SW complexes in the world, and when the last of the twelve ALLISS stations went on air in the autumn of 1996 RFI again possessed the most modern and advanced SW centre in the world.

RFI has a SW relay station at Montsinery in French Guiana, which is also maintained by TDF. It is fitted with three 500 kW SW transmitters and twelve fixed directive curtain arrays. More recently this station has been extended with two of the ALLISS concept, called Toucans. One is used by RFI as a relay and the other is used by the Swiss as a relay station for Swiss Radio International (SRI). According to TDF this second Toucan is a specially designed version with a frequency coverage from 9 to 21 MHz.

In Africa RFI uses four 500 kW SW transmitters and associated directive antennas at the Moyabi SW station in Gabon, which belong to the commercial broadcaster Africa No. 1. RFI also has exchange relay agreements with Japan NHK to use its 300 kW transmitter at Yamata and with the People's Republic of China for two transmitters of 120 kW power at sites in Beijing and Xian.

RFI also broadcasts a foreign service over MW transmitters in Toulouse (300 kW), Strasbourg (300 kW), Cyprus (600 kW) and Kunming in China (1000/500 kW). The relay in Cyprus is from the Radio Monte Carlo Middle East station at Cape Greco. This same transmitter also acts as a relay for Radio Canada International and the religious broadcaster Trans World Radio (TWR). RFI broadcasts is programmes over all these SW and MW stations in French and in the following additional languages: Arabic, Cambodian, Chinese, English, German, Lao, Farsi (Persian), Polish, Portuguese, Romanian, Russian, Serbo/Croat, Spanish and Vietnamese.

RFI commands a listening audience in excess of 80 million around the world, though even following the upgrading and expansion programme of 1996–1997 it is unlikely that RFI will surpass the 130 million listening audience of the BBC World Service. It is unlikely to happen because Britain – along with Voice of America – possesses a powerful advantage in that English is the more widely spoken language. But what RFI will have is a unique and unrivalled capacity to flexibly focus almost its entire SW transmission capacity to any particular region or country in the world. During the Gulf War other international broadcasters' inability to react swiftly to geopolitical change, and effectively target that region, was revealed.

TéléDiffusion de France (TDF)

TDF is the broadcast transmission subsidiary of state-owned France Telecom. TDF operates as a commercial enterprise and is the expert body on the provision of television and radio broadcasting and transmission facilities in France. It provides transmission services for the state broadcasters Radio France and RFI, plus similar services and facilities to the main commercial broadcasting and TV chains, as well as nearly half of the local radio stations in France. It also works in close collaboration with Deutsche Bundespost Telekom and Swiss Radio International (SRI). With some 11,500 transmitters and relays to maintain, TDF is almost certainly Europe's largest radio and TV transmission provider.

Note

The history of French broadcasting, together with accounts of Radio France International, TéléDiffusion de France and French state-owned broadcast manufacturing capabilities was covered in Chapter 29 of *History of international broadcasting* [2]. Appendix III of that book describes ALLISS low-profile transmitters.

Chapter 9
Voice of America

Voice of America was born in February 1942, in the dark days of World War II and its first broadcast was in German. Its mission was to combat enemy propaganda and to explain why the US was in the war. As the nation's voice to the world its purpose was to broadcast truth to its listeners, whether that news be good or bad, and in 1942 the news was bad, with American and allied losses everywhere. Nevertheless the Office of War Information (OWI) which had taken charge of VOA's hastily improvised transmitter network never flinched from reporting the truth. For as the first broadcast said: 'If we don't tell the truth when news is bad then surely they will not believe it when it is good.'

For the US Government to take charge of an international broadcasting network was a step into the unknown, for unlike Europe – where radio broadcasting was controlled by the state – America had not to that point followed that policy. By that time propaganda broadcasting had become a fact of life with Britain, Germany, Italy and Russia all having developed their skills. Much happened since then; World War II ended and a cold war with the Soviets began. In 1982 the National Security Council under the Reagan administration directed VOA to provide a stronger, more reliable signal into areas of the world important to US interests. To substantiate this it pointed out that other major international broadcasters, including republics of the Soviet Union, had many 500 kW SW transmitters, whilst VOA had none.

At a high point in the second phase of the cold war (1979–1986) Ronald Reagan gave his famous speech, 'Voice of America has been a strong voice for truth. Despite problems of antiquated equipment and Soviet jamming, the Voice of America has extended the message of truth around the world. Were it not for years of neglect that voice could be heard more clearly and that's why our administration has made the same kind of commitment to modernise the Voice of America that Kennedy brought to the space programme.'

Table 9.1 *Voice of America broadcasting transmitter stations, 1986–1987*

Relay Station	Primary programme feeds	No. of antennas	Total no. of transmitters (SW/MW)	VOA transmitters power (kW)	Total power (kW)	Age (years)	Target area
US Stations							
Bethany, OH	Leased tel. lines from Washington	22	6 SW	1 – 175 3 – 250 2 – 175	925	42 17 42	Latin America and West Africa (Used by AFRTS)
Delano, CA	Commercial satellite from Washington	15	9 SW	3 – 250 4 – 250 2 – 100	750 1000 200	17 1 41	Central America and East Africa (Used by AFRTS)
Dixon, CA	Commercial satellite from Washington	14	3 SW	3 – 250	750	17	Central America
Greenville, NC	VOA microwave from Washington	66	12 SW	6 – 500 6 – 250 4 – 500	4500 2000	32 22	Latin America, North and West Africa, and Western Europe
Marathon, FL	Commercial satellite from Washington	1	1 MW	1 – 50	50	23	Cuba
Total – US		118	30 SW 1 MW	26 SW 1 MW	9925 kW 50 kW		
Overseas Stations							
Judge Bay, Antigua	Commercial satellite from Washington	1	1 MW	1 – 50	50	17	Lesser Antilles down to St. Lucia
Bangkok, Thailand	Shortwave from Philippines	1	1 MW	1 – 1000	1000	32	Southeast Asia

Location	Source	No.	Transmitters	kW	Total kW	Freq.	Coverage
Selebi-Phikwe, Botswana	Shortwave from Greenville and Bethany	1	1 MW	1 – 50	50	4	Northern S. Africa and Southern Zimbabwe
Kavala, Greece	Commercial satellite from Washington	23	1 MW 10 SW	1 – 500 9 – 250	2750	32 14	MW – Eastern Europe (Romania, Yugoslavia) SW-Middle East, S. Asia, South Central and Western USSR and Eastern Europe
				1 – 250	250	14	(Used by Gov't of Greece)
Rhodes, Greece	Commercial satellite from Washington	8	1 MW 2 MW	1 – 500 2 – 50	600	32 22	Arabic Middle East
Monrovia, Liberia	Commercial satellite from Washington	27	8 SW	6 – 250 2 – 50	1600	21	Sub-Saharan Africa
Munich, Germany	Commercial satellite from Washington	17	1 MW 4 MW	1 – 300 4 – 60	540	39 49	MW – Eastern Europe (Czechoslovakia, Poland, Hungary); SW – Western USSR and Eastern Europe
Poro, Philippines	Commercial satellite from Washington	17	1 MW 4 SW	1 – 1000 1 – 35 2 – 100 3 – 50 1 – 35	1385 35	32 32 32 22 32	MW – Vietnam SW – China, Southeast Asia and E. Africa (Used by AFRTS)

Table 9.1 (*cont.*)

Relay Station	Primary programme feeds	No. of antennas	Total no. of transmitters (SW/MW)	VOA transmitters power (kW)	Total power (kW)	Age (years)	Target area
Tinang, Philippines	Commercial satellite from Washington	39	15 MW	3 – 50 10 – 250 2 – 250	3150	20 17 4	Eastern USSR, China, Southeast Asia and Eastern South Asia
Quesda, Costa Rica	Shortwave from Greenville	1	1 MW	1 – 50	50	17	Northern Costa Rica
Colombo, Sri Lanka	Shortwave from Kavala and Philippines	19	4 SW	2 – 35 1 – 10 1 – 35	80 35	32 32 32	India (Used by Gov't of Sri Lanka)
Tangier, Morocco	Commercial satellite from Washington	32	10 SW	3 – 100 4 – 35	440	35 40	Eastern Europe and North Africa
Putna Gorda, Belize	Shortwave from Greenville	2	2 MW	2 – 50	100	1	Northern regions of Central America
Woofferton, England	Commercial satellite from Washington	37	10 SW	6 – 250 4 – 300	2700	22 5	Western USSR. Caucasus and Eastern Europe
Total – Overseas			10 MW 70 SW	10 MW 64 SW	3550 kW 10945 kW		
		225	80	74	14495 kW		
Total – US & Overseas			11 MW 100 SW	11 MW 90 SW	3600 kW 18870 kW		
		343	111	101	22470 kW		

Reagan made this statement in March 1984, during a visit to Morocco to sign an agreement with the Moroccan Government permitting America to build a new SW station on their soil – perhaps the most ambitious project in VOA's history. His references to years of neglect had more to do with transmitter powers than the geopolitical infrastructure of VOA's global network, which relied on SW and MW transmitters located in 'host' countries, strategically positioned to be able to reach designated target countries or regions by ionospheric path lengths, and designed to achieve maximum audibility in the target zones. This strategic network ensured that the vast regions of the republics of the USSR, spanning eleven time zones, were within reach of VOA's SW transmissions.

The VOA worldwide network, as it existed in 1986–1987, its stations, locations, countries and numbers of transmitters and output powers, is shown in Table 9.1. Four 500 kw SW transmitters at the Greenville station have been added to the list because these were supplied during 1987 for evaluation as part of the RFP tender for the VOA Modernisation Programme. These 500 kW transmitters, of the type manufactured by Continental, AEG, Brown Boveri, Marconi and Thomson, were to form the backbone of VOA's new transmitter network.

The introduction of such transmitters, with their high initial cost and higher operating costs, gave some impetus to the development of more powerful directive curtain arrays of the 4 wide, 6 high type, with slewing facility in both azimuth and elevation, because ultimately it is the antenna which determines the performance of the entire system.[1]

Modernisation of VOA's global network

In 1982 the National Security Council directed VOA to provide a stronger, more reliable signal into areas of the world where the United States had important interests. The chief reason for this was that at that time the outcome of the cold war was far from certain, and US international broadcasting from VOA and RFE/RL was in some regions outclassed by the performance of the Soviets, who had some 200 kW SW and nearly 100 high power MW transmitters at their disposal.

In fiscal year 1983 a formal modernisation programme began at VOA, with recruitment of skilled engineering staff, and detailed studies and plans for an updated SW relay network. Never a nation to do things on a small scale, when finally the 'request for price' tender documents were issued to the major transmitter manufacturers a quantity of 100 500 kW SW transmitters was called for, though eventually this quantity was reduced to 55 – still a very large order.[2] The overall project time for the planning and

[1] For more on the history of VOA, see *History of international broadcasting* [2] and references [5, 6].
[2] See *History of international broadcasting* [2], pp. 167–193, for further details of this tender.

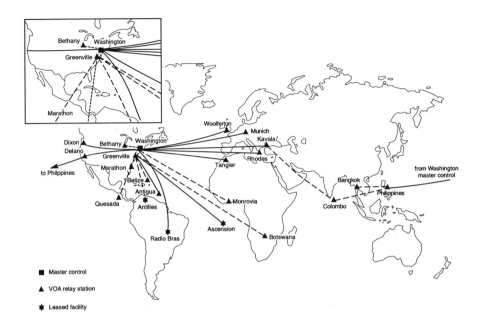

Figure 9.1   *Voice of America: transmission methods to relay stations*
——— *satellite*, – – – *shortwave*, - - - - - *microwave or landline*

construction of a single high power, multi-frequency SW transmitter station can be several years. In the case of VOA's modernisation programme, which was of a global nature, the project times were further lengthened due to the legal and diplomatic negotiations with 'host' countries. The best example of this is that of the SW station planned in Morocco. This project commenced officially in March 1984, though planning had started a year before. Even so, the new station was not completed until summer 1993 and the station was finally dedicated on 4th October, 1993.

But with a project of this size other factors came into play; these were to do with fiscal realities and shifts in world politics in regions of the world beyond the FSU and Eastern Europe, such as the Middle East and SE Asia where US foreign policies were also being pursued. Such factors made it necessary for the USIA and the Bureau of Broadcasting to reassess and fine tune the VOA modernisation programme from time to time. Satisfying mission requirements involves a study of factors such as defining target areas, distance to target, estimates of the listening audience and consideration of the competition – what other international broadcasters are doing.

Table 9.2 shows the state of the VOA modernisation programme by January 1994, although by 1995 the programme had moved a few more

Table 9.2 *Voice of America existing and planned broadcast transmitter network, January 1994*

Current relay stations	Manufacturer/ model	Broadcast transmitters		Total power	Age (years)	Comments
Belize	Harris VP-100B	2 – 100 kW	MW	200 kW	7	To be closed March 1994
Bethany, Ohio	Collins 821A-1	3 – 250 kW	SW	1500 kw	32	SW antennas to be
	ABB SK53C3	3 – 250 kW	SW		14	replaced end of FY96
Botswana, Selebi-Phikwe and	Continental 317C-2	1 – 50 kW	MW	50 kW	14	To be transferred to govt. of Botswana in 1995
Moepeng Hill	Continental 418E	4 – 100 kW	SW	400 kW	3	1 – HP MW planned
Colombo, Sri Lanka	Collins 207B-1	2 – 35 kW	SW	115 kW	44	3 – 500 kW SW and
	Philips	1 – 10 kW	SW		42	1 – 500 kW SW*
						planned for new site
	Collins 207B-1	1 – 35 kW	SW		44	Used by host govt.
Delano, California	Collins 821A-1	3 – 250 kW	SW	1750 kW	32	
	BBC SK53C3	4 – 250 kW	SW		14	
Greenville, North Carolina	Continental 420A	6 – 500 kW	SW		42	6 – 500 kW SW planned*
	Continental 420B	1 – 500 kW	SW		9	
	GE 4BT250A1	6 – 250 kW	SW	6500 kW	34	6 – 500 kW SW planned*
	Marconi B6127	1 – 500 kW	SW		9	
	BBC SK55 C3-2P	1 – 500 kW	SW		9	
	AEG S4005	1 – 500 kW	SW		9	
Kavala, Greece	Continental 105B	1 – 500 kW	MW	3000 kW	42	1 – HP MW planned
	Continental 419D	9 – 250 kW	SW		25	9 – 500 kW SW planned*
	Continental 419D	1 – 250 kW	SW		25	1 – 500 kW SW planned*
						(used by host govt.)
Kuwait	Continental 317C-2	1 – 100 kW	MW	100 kW	14	Interim: 1 – 600 kW planned
Morocco	Marconi B6-128	10 – 500 kW	SW	5000 kW	1	

Location	Transmitter	Quantity – Power		Total	No.	Notes
Munich	RCA BTA-150	1 – 300 kW	MW	700 kW	46	1 – 300 kW MW planned
and	Radio Slavia K.V. 100/B	2 – 100 kW	SW		54	
Wertachtal, Germany	Radio Slavia S.W.B9 R/B	2 – 100 kW	SW		54	Not in service
	AEG S4005	2 – 500 kW	SW	2000 kW	6	Leased from Deutsche
	AEG S4005	2 – 500 kW	SW		7	Bundespost
Philippines, Poro	Continental 105B	1 – 1000 kW	MW	1420 kW	42	1 – HP MW planned*
	General Electric 100C	2 – 100 kW	SW		41	4 – 100 kW SW planned*
	Collins 207B	2 – 35 kW	SW		44	
	Gates HF-50C	3 – 50 kW	SW		34	
and Tinang	Hughes HC-114	10 – 250 kW	SW	3150 kW	27	12 – 50 kW SW planned*
	Gates HF-50C	3 – 50 kW	SW		34	
	BBC SK53C3	2 – 250 kW	SW		14	
Rhodes, Greece	Continental 105B	1 – 500 kW	MW	600 kW	42	1 – HP MW planned*
	Gates HF-50	2 – 50 kW	SW		34	
São Tomé	Harris VP100B	1 – 100 kW	MW	100 kW	7	4 – 100 kW SW planned 1 – HP MW planned
Thailand, Bangkok	Continental 105B	1 – 1000 kW	MW	1000 kW	42	1 – HP MW planned 7 – 500 kW SW on air in
and Udorn	Marconi B6-128					1994 (one for host govt. use)
Woofferton, England	Marconi BE124	4 – 300 kW	SW	1700 kW	14	Leased from BBC
	Marconi BD272	2 – 250 kW	SW		34	Leased from BBC
Totals		10 MW 95 SW		29285 kW	25	
Additional planned relay stations						
Pacific Islands*		4 – 500 kW	SW	200 kW		

Notes:
* Project not approved by OMB or Congress
HP – high power
Source: Voice of America

Figure 9.2   *Interior of VOA transmitter building at Morocco SW station*

steps to full completion, with new and more powerful SW stations coming
on stream, the first of the new generation being in Morocco.

Morocco

When it was completed in 1995 this SW station was the culmination of
more than a decade of planning which went back to the Reagan administration days and the commitment to modernise VOA. Located at the Briech
site in Morocco on the shores of the Mediterranean, this SW facility might
be termed the flagship of the entire VOA global network. It enables the
'Voice' to reach out across several thousand kilometres to Eastern Europe,
the Middle East and to parts of SE Asia.

Equipped with ten B6128 500 kW SW transmitters supplied by Marconi
under a separate equipment supply contract, a 10 × 23 coaxial switching
matrix connected by coaxial cable to 21 HR 4/6 curtain arrays supplied by
TCI of Sunnyvale, California, steerable in both azimuth and elevation
angles, this SW station is able to target precisely many different regions of
the world.

Because these curtain arrays are built out on what was once the sea,
the propagation characteristics for angles of elevation extending almost to
grazing are near-perfect; there are no obstructions in the forward plane.
However the project entailed moving 10 million cubic feet of earth and

Figure 9.3 *One of the SW curtain arrays supplied by TCI to VOA SW station in Morocco*

much piling. From a civil engineering standpoint this SW site is the most ambitious project ever undertaken by the US Corps of Engineers, and because an open-wire feeder was not practical some 16,000 m of Andrew's Semi-Flexible cable was used between the transmitter matrix and the curtain arrays. Although the final cost is not available it would be surprising if it came to less than US$ 200 million.

Thailand

This SW station, completed in 1995, is similar in concept to the Morocco station but has seven 500 kW SW transmitters and 25 curtains similar to those at Morocco (supplied by the same contractor, TCI). In both stations the installed curtains were of the latest technology; fourth generation 4 wide/6 high, developed especially by TCI for the VOA modernisation programme. The Thailand facility reaches out to fill gaps in the old VOA network and provide better coverage to China, SE Asia and parts of the former USSR.

Kuwait

The Gulf War gave the US the opportunity to establish a station in Kuwait, using MW to reach out to adjoining states, Afghanistan and Iran. This facility was started up with two 50 kW transmitters and combiners, but the Bureau for International Broadcasting then successfully negotiated a deal with Kuwait which permitted the siting of a much higher power facility. This agreement, signed on 2nd August 1992, enabled the Bureau to construct a 600 kW MW facility using a Marconi B6043 600 kW transmitter and an antenna system designed by TCI. The redeveloped station came on air in May 1996 and not surprisingly brought a hostile response from the Islamic republic of Iran.

Sri Lanka

A 1989 visit by VOA engineers identified Sri Lanka as an ideal location to fill some critical gaps in broadcast coverage to China, SE Asia and the Middle East. Evidently the USIA, the Bureau of International Broadcasting and VOA all regarded it as important for a SW facility to be operational here in the shortest possible time, because for the first time in the history of VOA it awarded the entire contract to a single contractor with the objective of swift completion. GEC-Marconi won the much-contested VOA tender for the 500 kW transmitters and was awarded the contract. The station, completed in 1996, was almost immediately destroyed by a transmitter fire, requiring rebuilding work. This station is equipped with four Marconi type B6128 600 kW SW transmitters, fourteen high performance curtain arrays – four of which are of the 4 wide/4 high Type HRS 4/4/.5, whilst the remainder are 6 high, Type HRS 4/6/.5. This is believed to be the first time Marconi has built the latter.

São Tomé and Principe

To compensate for the loss of the Liberia relay facility in September 1990, it was essential that VOA should regain control of Sub-Saharan Africa. The Bureau identified the island of São Tomé in the Democratic Republic of São Tomé and Principe, in the Gulf of Guinea as the site for a new relay station. In the spring of 1993 the Bureau completed an interim facility of a 100 kW MW station using a Harris-built VP 100B. This then permitted VOA to reach the area extending from the west coast of Africa up to the Ivory Coast. In July 1993 a design-and-build contract was awarded for four 100 kW SW transmitters. Won by the Swiss company ABB Infocom – now part of Thomcast – the Type SK51 C3 3P6 100 kW transmitters were delivered in 1995.

Botswana

A further result of the loss of the Liberia Relay facility was the rapid deployment of a SW station at Moepeng Hill in northeast Botswana. The first two systems, comprising generators, transmitters and antennas, were completed in December 1991 by the US company TCI, which also supplied the eight curtain arrays, Type HR 2/4/.5. The final two 100 kW systems were completed in March 1992, in time for when US forces were deployed in Somalia.

Figure 9.4   *View of transmitter buildings and antennas at São Tomé*

Because the Bureau of Broadcasting had previously agreed to turn over its existing 50 kW MW facility at Selebi-Phikwe to the government of Botswana in 1995, a new facility for VOA was required in Botswana. This facility was also constructed by TCI and used a Marconi transmitters Type B6044, which had been supplied against the VOA contract.

Philippines

This facility in Poro is one of VOA's oldest stations. Before renovation its newest transmitters were approaching 40 years of age, whilst its 1000 kW MW transmitter was built in 1952. It was retired from service in 1995 when it was replaced by a Harris DX 1000, a one MW medium wave transmitter. Elsewhere in the Philippines, at Tinang, VOA has a SW facility, now over-due for new 500 kW SW transmitters to replace ten 250 kW transmitters built by the Hughes Corporation in the 1960s. Transmitters such as these underline the serious equipment problems which face VOA staff, which in some cases mean that spares have to be specially manufactured.

Thailand

The Bangkok facility's 1000 kW MW transmitter, installed in 1952, was eventually replaced in 1996 with a DX 1000 supplied by Harris, giving the station a much higher performance in electrical efficiency and in depth of modulation, and a correspondingly larger service area.

Munich, Germany

The Munich facility is one of VOA's oldest stations. The original 300 kW MW transmitter, manufactured by RCA in 1950, was replaced with a 300 kW solid state version from Thomson-CSF in 1994.

Belize

This tiny Central American country, located in the southeast corner of the Yucatan peninsula of Mexico, is an example of the strategic value of an otherwise minor country. Less than 600 km from Cuba, the VOA station at Punte Gorda had a high strategic importance for many years, and its two 100 kW MW transmitters were declared of no further use to VOA after less than eighteen months, due to the changes that had occurred in the Latin American political climate and also by political reform in the former Soviet bloc which effectively removed the threat of communism in Central America. Thus Belize was officially shut down in March 1994.

Bethany, USA

Bethany is one of the VOA's older stations and in 1989 its three 45 year old transmitters were taken out of service and replaced with three 250 kW ABB SK53C3 transmitters.

Delano, USA

This SW site in the VOA network was selected to carry out research and development into more powerful curtain arrays. The 12 by 6 curtain array, constructed in 1986, makes Delano the most powerful SW directive antenna in the western hemisphere, with a forward gain of 30 dB. More detail about this site can be found in Chapter 24.

Greenville, USA

This facility in North Carolina is the largest station in VOA's network in Continental America. Greenville plays a key role as the test bed for new transmitters. It is scheduled to have twelve new 500 kW SW transmitters but this project was still awaiting fiscal approval at the time of publication.

Israel

During the final decade of the cold war the construction of a major new SW station in Israel, part of the VOA modernisation programme, was planned as a joint development between the VOA and the Board for International Broadcasting. The project suffered delays due to environmental concerns and associated legal actions in Israel, although whether these factors were the only reasons is a matter of conjecture. By 1990 the threat from the cold war had gone. The project was cancelled in July 1993. There is no doubt that it could have played a key role in the cold war had it been constructed in the late 1980s.

The VOA BSE research programme

In 1984 as part of the VOA modernisation programme, a research programme was initiated to provide data for antenna procurement specifications. The need to overcome jamming signals amongst other factors, had focused attention on the need for more efficient antennas for SW broadcasting.

Broadcast Systems Engineering (BSE) contracts for the research were awarded to a number of companies, which included Holmes and Narver as Prime Contractor, Atlantic Research Corporation, International Broadcasting Consultants and National Teleconsultants. The BSE contract was awarded by the US Army Corps of Engineers, who provided guidance and assistance. In 1985 a further contract was awarded to TCI to develop a prototype for a state-of-the-art antenna for SW broadcasting.

Studies conducted by the BSE Group provided substantial data on vertical angle propagation for a wide range of target azimuths and distance ranges. VOA requirements cover a wide range of targets ranging from small countries to substantial parts of continents, with distances that range from about 2000 to many thousands of kilometres. The selection of antennas is thus complicated because they need to service small, medium and long range targets. Some of the most significant findings of the BSE Research programme are given here.

A propagation study for 43 specified language areas of the world showed that 6 stack curtain arrays proved best for 23 areas (54 per cent); 4 stack was best for 19 areas (44 per cent) and 2 stack was best for only one area (2 per cent). A comparison of performance between 4 stack and 6 stack showed that generally the 6 stack provides a signal about 2 dB better. It was further established that vertical take off angles (TOA) ranged from 2–22 degrees and were fairly independent of operating frequency.

It was realised that selection of the best available vertical and horizontal patterns for a given target area as a function of frequency provides an effective method for offsetting the substantial variations of radiation patterns over the octave bandwidth of the antenna. This technique requires multiple beam selection by switching excitation of the dipole radiators in each bay of the antenna with limited control of horizontal apertures. Lowest vertical angles are obtained in the +++ mode, that is when all stacks are energised. The ++ mode was next best – when the two upper stacks are energised. The ++ mode is equivalent to an HR 4/4 and gives lower gain than the +++

Figure 9.5   *VOA relay station at Tinang, Philippines*

mode at low vertical angles, and higher gains at higher vertical angles. In brief, then, this BSE research study proved the superiority of the HR 4/6 for maximum target penetration for one or two hops, between 4000 and 9000 kilometres.

The VOA research programme was one of the most exhaustive studies ever carried out in this branch of radio science. It was paid for with American taxpayers' money and is a good example of how wars – hot or cold – act as a spur to the development of technology.

VOA programme placement

A modern VOA facility for central network control and management of relay stations and programme feeds operates on a 24 hour basis. This Network Control Center (NCC) routes Washington-produced programmes to VOA relay stations around the world, maintains continuous monitoring and display of the status of all broadcast-related equipment in the global network, and assists the relay stations in accomplishing their broadcast missions.

Satellite communications have played an increasingly important role in the technical and engineering operations of the International Broadcasting Bureau, and the latest generation of geo-stationary satellites are now used in the VOA global network.

Satellite Interconnect System (SIS)

Two-way earth stations are now in operation in Washington DC, Bethany, Botswana, Morocco, São Tomé, Tinang and Udorn. Gateway SIS earth stations have been installed in Greenville, Delano, Poro and Munich. These earth stations allow programmes received from one satellite to be re-transmitted to another satellite in the various ocean regions, whilst receive-only SIS earth stations are operational in Bangkok, Botswana, Kuwait and the UK.

Specific developments which have now come on stream include two-way Indian-Ocean (IOR) gateway earth stations in Poro and Tinang, two-way Pacific Ocean (POR) region earth stations also in Poro and Tinang, an IOR gateway earth station and a two-way Atlantic Ocean (AOR) relay station in Munich, a two-way AOR earth station in São Tomé, receive-only AOR earth stations in the UK and São Tomé and a receive-only IOR earth station in Kuwait.

Technical monitoring

The International Broadcasting Bureau also operates seven Technical Monitoring Offices (TMOs) at strategic locations around the world. These monitor the technical quality at various points in the VOA broadcast chain.

Figure 9.6   *A VOA main gateway SIS earth station at Delano, CA, USA*

Chapter 10
Radio Canada International

As RCI prepared to celebrate its 50th anniversary in 1995, there was a great deal of uncertainty about its future. The previous two or three years had seen one government department after another coming in for across-the-board cuts in operating budgets, brought about by the recession. In December 1955 it was RCI's turn to receive what seemed a fatal blow – it was to cease all operations on 31st March 1996.

That RCI is still alive is due to a worldwide effort by RCI's staff, friends and listeners, who banded together to fight what seemed to be inevitable. Thousands of letters were delivered to the Prime Minister's office as well as to other cabinet ministers. There was also support from Canada's other broadcasters, public, private and commercial, as well as from the printed media. All of these campaigns, spontaneous and uncoordinated, reached a climax and a victory when the Deputy Prime Minister Sheila Copps indicated (at what must have seemed the eleventh hour) that the Voice of Canada must not die, and that funds would be found to keep RCI on the air during 1996 while a permanent solution to the budget question was found.

This last minute reprieve was testimony to the efforts of thousands of people who fought for the survival of one of the world's most respected international broadcasters. No one was happier than Jacques Bouliane, Director of Engineering who in 1996 said 'Leave it to us to do everything possible to make sure this one-year period gets extended indefinitely'.

Measured by its numbers of transmitters and total kW output power, Radio Canada International does not measure up to the big broadcasters such as the BBC World Service, Deutsche Welle, Radio France International or Voice of America. But what RCI lacks in kW power is compensated for by its influence in the world. It played a very important part in the cold war and the 1980s saw Radio Canada International consolidate its position as one of the world's most trusted broadcasters.

The idea of Canada having an international voice of the short waves was first proposed as far back as the 1930s. Studies commissioned by the Canadian Broadcasting Corporation came to this conclusion and the need was recognised by Parliamentary Broadcasting Committees. In 1942 it was announced by the Prime Minister that Canada would begin a SW service and the CBC International Service became a reality on 18th September 1942. Even so it was to be another two and a half years before the first test broadcasts were made to Canadian troops in Europe on 25th December 1944. This was the time it took to locate a studio for the new International Service and to conclude the search for a suitable SW transmitter site, for which after a careful study, Sackville, New Brunswick was chosen. During the latter part of 1943 two 50 kW RCA SW transmitters were installed along with directive antennas.

As a SW transmitter site Sackville proved to be a near-perfect choice and it provided a consistently good circuit to Britain and mainland Europe. Throughout the war years the International Service of CBC concentrated its broadcasts eastwards to Europe, and south to the Caribbean and Latin America. In the aftermath of World War II and with the onset of the cold war, CBC was drawn into the cold war to play its part in the broadcasts from the West to the countries behind the iron curtain. CBC's International Service (IS) began beaming SW broadcasts to Czechoslovakia in the Czech and Slovak languages in 1949 and a Russian language service followed in January 1951. This was followed by a Ukrainian service in September 1952 and a Polish service one year after that in 1953. The Canadian immigrants who took part in these broadcasts were sometimes dissidents and defectors from these countries and were able to identify with their countrymen and let them know what was happening in the world. With the deepening of the cold war, transmissions to the countries of Eastern Europe and the Soviet union intensified throughout the 1950s. Canada was now becoming an important player in the war of words against communism. In 1962 the CBC service acquired a third 50 kW SW transmitter to give increased coverage to Europe and Africa. It should be remembered that 50 kW was a high power transmitter at that time.

Though few listeners were initially aware of the fact, and despite being operated by CBC from 1945, the International Service (IS) was a separate entity controlled by the parliament of Canada, owned and funded by the Department of External Affairs. In July 1970 the International Service was renamed Radio Canada International (RCI), bringing it into line with the title used by several other countries, such as Radio France International.

In recognition of the fact that the new RCI had fewer transmitters at its disposal compared with other Western broadcasters, RCI capital project grants enabled it to acquire three new Collins 250 kW SW transmitters in 1971 and 1972, five times more powerful than the existing transmitters, enabling it to increase its signal audibility to Europe, the USSR and Africa. Shortly after this RCI concluded an agreement with the British Broadcasting Corporation to acquire and operate two high-powered SW transmitters at the Daventry transmitter station. This was RCI's first venture into owning and operating transmitters away from Canadian soil. A SW link between RCI's Montreal studios and Britain enabled broadcasts from Canada to be relayed via Daventry so as to give a one-hop ionospheric path to countries in Eastern Europe and to parts of the Soviet Union.

That experience with a SW relay using satellites led RCI to consider using more such relays to bridge the distance between it and the target zones, and in 1972 RCI made a similar agreement with Deutsche Welle to use its SW facility at Sines in Portugal. In return, RCI gave Deutsche Welle an equal amount of air-time over its Sackville facility, thus enabling the German broadcaster to project a stronger signal into the Caribbean and to North America. The Sines facility, located on the west coast of Portugal, gave RCI the ideal ionospheric path to most of Eastern Europe.

At the end of 1979 RCI was broadcasting 27 hours of programming daily to North America, South America, Africa, the Middle East, Western Europe, Eastern Europe and the USSR in eleven different languages (English, French, German, Spanish, Portuguese, Russian, Ukrainian, Polish, Czech, Slovak and Hungarian). In 1985 it acquired three 100 kW SW transmitters from Harris to replace the older 50 kW RCA models.

Throughout its first 39 years RCI had been beaming its SW broadcasts eastwards, and in 1984 it took the decision to look westwards for new audiences in Japan. RCI and Radio Tanpa of Chiba entered into an agreement which permitted RCI to begin regular broadcasts in Japanese, and the first programme went on air in May 1984. Five years later a Chinese language service began. On 1st October 1989 RCI began a direct service to listeners in China using the SW transmitter complex at Yamata, owned by NHK.

Relay agreements

By this time exchange relay agreements had become so commonplace that SW listeners could no longer be sure where a broadcast was coming from. All the major Western broadcasters had concluded bilateral agreements of one kind or another making it possible to reach listeners in the target zones with much stronger signals. RCI concluded some highly satisfactory relay exchange agreements, which by 1990 included those listed in Table 10.1.

Table 10.1 *RCI relay broadcasting agreements*

Broadcaster	Country	SW transmitter site
BBC World Service	England UK	Daventry
Deutsche Welle	Federal Germany	Bavaria
Radio Austria International	Austria	Moonsbrunn
Radio Korea International	Korea	Kimjae
China Radio International	China	Xian
Radio Monte Carlo	Monaco	Cape Greco, Cyprus

Radio Canada International and the cold war

Radio Canada International (RCI) was one of the international broad-casters which found itself in the propaganda maelstrom after it became evident to Britain and America that RCI could play a part in their propaganda broadcasting, a reasoning evidently based around the fact that Canada had a large number of emigrants from East European countries such as Czechoslovakia, Poland, Russia and the Ukraine, some of whom were refugees or dissidents. Such a cross-section of people could provide the basis for news and propaganda programmes for listeners behind the Iron Curtain. Moreover some of the newer dissidents were able to expose the corruption that went on behind the Iron Curtain. Being a dissident could get them a job and, the more important the dissident, the greater the salary. Thus by the mid 1950s RCI was playing a role in the cold war.

As evidence that promoting peace and understanding to nations was never a front runner in international broadcasting, the language broadcasts from RCI throw up some interesting facts. Before the onset of the cold war RCI was primarily concerned with broadcasting in the English and French languages along with a few others such as Dutch, Danish, Italian, Norwegian and Swedish. As the cold war intensified so the Canadian Broadcasting Corporation decided that 'broadcasts to Western Europe were not as important as they had been previously' and 'that it was more important to concentrate on broadcasting to Eastern Europe' [8].

Accordingly, on 4th March 1961 the Danish, Dutch, Italian, Norwegian and Swedish services were all discontinued to make way for language broadcasts to the USSR and Eastern European services. These broadcasts were obviously seen as a danger by the Soviets because from the mid 1950s many SW listeners in North America can recall Radio Canada becoming swamped by the Soviet jamming which occurred when RCI's English programme switched to Russian, Polish or Czech. For the Russian jammers to be audible in North America signifies the use of sky-wave jammer transmitters of enormous carrier power.

Transmissions from the Sackville SW station in New Brunswick were always heard with good audibility in Western Europe because of the nearly all-sea path, but throughout the late 1950s and the early 1960s RCI began a new technique aimed to combat Soviet jamming. Sackville transmissions were picked up in Britain by a SW receiver station, recorded on tape and then re-broadcast via 50 kW SW transmitters owned by the BBC at Daventry.

The objective behind the strategy of using another SW station nearer to the targeted country was of course to convert a 2-hop ionospheric path from Sackville, Canada (to the target zones) into a 1-hop path from the relay station, thereby improving signal audibility in the USSR and Eastern Europe. Where such relays were used these should have been published in the broadcast schedules of RCI, but the first relay station at Daventry was never disclosed. However, the increased audibility of Radio Canada in the target zones made the use of relays obvious to the Soviet authorities. Accordingly in March 1967 RCI began to list the relay in its broadcast schedules.

In 1963 The Radio Canada Shortwave Club was created. The purpose of such clubs is to stimulate listening interest in target countries. Members may become suitable for political indoctrination without being aware of it (the author, age 16, was invited and became a member of Moscow Radio Shortwave Listeners Club in 1936, bringing himself quite unknowingly to the attention of the British government as a suspected communist). To what extent the Canadian broadcasts were successful in recruiting SW listeners is open to speculation, but since SW listening was a popular hobby in Eastern Europe it is likely that many became regular listeners.

In 1967 RCI purchased two of the Daventry transmitters from the BBC. With the aid of satellite communications, which by this time were in regular use, RCI was able to fully integrate these ex-BBC transmitters into its own network, thus permitting live broadcasts between RCI's Montreal studios and the SW relay station at Daventry.

In 1971 RCI concluded an agreement with Deutsche Welle to use its SW facility at Sines, Portugal. The choice of this SW transmitter facility provided further optimum distance-to-target routes to reach East Germany, Poland, Czechoslovakia, the Ukraine and Russia, especially in early mornings and late afternoons.

By 1979, at the height of the cold war, RCI was penetrating the iron curtain with SW broadcasts direct from Sackville, as well as its SW facilities in Europe. BBC Daventry was relaying transmissions received via the mid-Atlantic ocean relay satellite from Montreal, whilst Sines was able to pick up off-air RCI transmissions from Daventry and signals direct from Sackville for re-broadcast by SW.

With the end of the cold war came a reappraisal of the role of RCI. In early 1991, with the Canadian economy still not recovered from the 1980s recession, the government cut back its budget. Six of the thirteen language services (Czech, German, Polish, Hungarian, Portuguese and Japanese) were cut. Further budget cuts almost proved terminal for RCI but in the end wisdom, public opinion and common sense prevailed, and now RCI looks forward to a more secure future.

Table 10.2 *Key milestones in RCI history*

1942	Prime Minister of Canada signs an order-in-council for Canada to have its own international radio service
1943	Studios built and a SW transmitter site selected at Sackville, New Brunswick, fitted with two 50 kW RCA transmitters
1945	CBC International Service goes on air in French and English
1949	Language services expanded to include Czech, Dutch, followed by Swedish, Danish, Norwegian, Spanish and Portuguese
1951	Russian service is added
1954	The service provides 16 hours of daily programming, but cuts made in several languages so as to concentrate more on the cold war
1961	Danish, Dutch, Italian, Norwegian and Swedish discontinued to allow concentration on cold war; more emphasis on East Germany
1963	CBC forms a Shortwave Listeners Club
1963	CBC re-broadcasting its transmissions from BBC Daventry
1970	CBC International Service renamed Radio Canada International
1971	RCI acquires two high power SW transmitters at Daventry, England
1971	RCI acquires three new Collins 250 kW SW transmitters for Sackville
1972	RCI concludes agreement with Deutsche Welle to relay programmes from Sines, Portugal
1973	RCI concludes relay agreement with Deutsche Welle to use Cyclops facility, Malta
1979	RCI broadcasting 27 hours daily programming in 11 languages
1984	RCI concludes an agreement with Japan NHK to re-broadcast RCI programmes
1989	RCI broadcasting to China from Yamata, Japan
1991	Budget cuts compel cut-backs in RCI staffing and withdrawal of 6 languages
1995	Complete closure and shut down of RCI announced for March 1996
1996	RCI gets a one-year reprieve from closure (subsequently extended)

Chapter 11
Swiss Radio International

On 1st August 1934, the Swiss Broadcasting Corporation (SBC) began experimental SW transmissions using the League of Nations SW transmitter site in Prangins, near Geneva. Reports of these broadcasts from as far away as South America prompted SBC to introduce a monthly, one hour service, starting up in the autumn of 1935 for Swiss living in North and South America. These broadcasts triggered considerable interest from not only Swiss, but also non-Swiss listeners and prompted SBC to broadcast weekly instead of monthly. At the same time SBC submitted an application to the Swiss Post, Telephone and Telegraph (PTT) to operate its own SW transmitter.

With a transmitting licence subsequently granted, and an initial budget granted by Parliament, SBC set about the construction of its own SW transmitter site at Schwarzenburg, and on 1st July 1939 a 25 kW transmitter was commissioned into regular SW service. Unfortunately disaster struck exactly one week later when the transmitter station was destroyed by fire. Once again SBC turned to the League of Nations for the use of its SW site at Prangins. Subsequently, in the autumn of 1940, one year into World War II, SBC inaugurated a regular SW service from its re-built transmitter station at Schwarzenburg.

World War II contributed to Swiss Radio International's rapid growth as an international broadcaster and to its emergence as a reliable source of news and information. As the voice of neutral Switzerland, SRI's untainted and objective news reports on the progress of the war in Europe began to attract the attention of SW listeners and governments around the world. For millions of people its broadcasts were virtually the only way to obtain objective news and a realistic picture of events in the war. By the time World War II ended Swiss Radio International was broadcasting in English, Spanish and Portuguese in addition to the three Swiss languages of French, German and Italian.

For a number of decades SRI programmes were dominated by its daily 'rendez-vous' broadcasts on SW, and from 1970 to the early 1990s, thirty minutes was the more or less standard length for all its programmes, which comprised international and home news, commentaries, press reviews, interviews and reports; whilst at the weekends the emphasis switched to entertainment. Since the end of the cold war SRI has, like some other international broadcasters, adopted a more flexible and constructive response to the changed situation, using its full compliment of programme and technical resources to ensure a more effective presence abroad. SRI has entered into partnerships with other organisations in Switzerland and with other media, while at the same time preserving its journalistic and editorial independence.

As the voice of neutral Switzerland, which has no political ambitions in regions of the world in conflict, SRI is very aware that crises in regions of the world where major powers do have political interests can influence the way international news is presented by the major international broadcasters. In this context SRI has a political independence which is probably greater than any other international broadcasting agency, and it has few constraints on its selection and diffusion of international news, comment, opinions and analysis of events.

Following the break up of the USSR, SRI was one of the first international broadcasters to use the opportunity to open up a dialogue with the newly liberated Eastern bloc countries. Media representatives from these countries were invited to Switzerland to take up residence for two months and given complete freedom to interpret Swiss facts and events and to broadcast to their own country. As a result of this first exercise, SRI became able to advise countries on how to start up an independent, free media – which, in reality, could take them many years to achieve in full.

Programme delivery

SRI has switched from exclusively using SW to a dual delivery via SW and satellite. In 1994 it launched the first two of four round-the-clock satellite programmes, one in French via EUTELSAT and the other in English via an ASTRA satellite, which can be received direct in millions of homes in Europe, and also via cable heads which serve many towns and cities in Western Europe. German and Italian satellite services were started up one year later. All SRI programmes will gradually be beamed by satellite to non-European countries via additional satellite channels.

Until portable and mobile satellite receivers are as freely available and inexpensive as the cheap SW receivers in common use in parts of the Middle East, Africa and large parts of Asia, SRI believes SW broadcasting will

Table 11.1 *SRI SW transmission capacity, 1997*

Site	Transmitter power	Target zone
Lenk	2 × 250 kW	Europe
Schwarzenburg	4 × 250 kW, 1 × 100 kW	Beyond Europe
Sottens	1 × 500 kW	Beyond Europe
China	2 × 120 kW	Asia and Middle East
French Guiana	1 ×500 kW	North and South America, also the Caribbean and SE Asia

remain indispensable for one or two more decades. Indeed, the Swiss Broadcasting Authority believes that delivery by SW might need to be reinforced with more powerful transmitters to some target zones in Africa and Asia.

Until the 1980s Swiss SW broadcasts were made exclusively from Swiss soil, from transmitter sites at Beromunster, Lenk, Samen, Schwarzenburg and Sottens, using mostly transmitters from the Brown Boveri Company of 100 and 250 kW power. In the 1980s the major international broadcasters like the BBC World Service, Deutsche Welle and Voice of America began replacing or upgrading their former 250 kW transmitters with newer and more powerful models of 500 kW carrier power. For environmental reasons the plans of SRI to follow this trend were foiled by residents living in the vicinity of its transmitter sites. It was this that led SRI to secure relay agreements in other parts of the world, much as described for RCI in Chapter 10. As a result, Swiss SW programmes are relayed from two 120 kW SW transmitters in the People's Republic of China and a 500 kW SW transmitter, with a rotatable curtain array, in Montsinery, French Guiana. Table 11.1 shows its SW transmission capacity.

With a total SW broadcasting power of 2,840 kW, Swiss Radio International is a long way down the world league table in terms of power. Yet measured by other standards such as credibility, impartiality and objectivity, the Swiss international SW service emerges as one of the world's most important international broadcasters. According to audience feedback, its impartiality is the main reason why listeners – in both crisis areas and other regions such as North America – prefer to tune into international news from SRI, geared to a brief daily contact with target audiences on all continents.

A listener survey in 1995 showed a SRI postbag mail figure from the Arab world of 3,657 letters, only slightly less than that for the whole of Western Europe, which is an indication of the influence SRI exerts in that part of the world.

Table 11.2 *SRI budget for year 1996*

Budget	SFr
Funds allocated by SBC	15,470,000
Contribution from Federal government	12,700,000
Other incidental revenue	1,585,000
Total	29,755,000

Funding

SRI is probably alone in Europe as an international broadcaster not wholly funded by the state. Under the Swiss Radio and Television Act, at least half of SRI's costs must be financed by the Federal government with the balance coming from the Swiss Broadcasting Corporation (SBC) licence revenue. In practice the Swiss government subscribes only three fifths of the total operating costs (Table 11.2), and SRI funding has long been a controversial issue as the SBC has always maintained that broadcasting services which promote Swiss foreign policy interests should not be financed out of licence revenue.

In making its assertion that international broadcasting should be financed by the state, SRI is transparently more honest than some other international broadcasters. The BBC World Service, for one, is fully funded by the UK Foreign and Commonwealth Office, yet it persists with its claim that it does not broadcast in the interests of the state but is the impartial voice of the British Broadcasting Corporation.

The total operating costs for SRI staff, programme production and transmission are estimated at SFr 29,827,000 for the year studied (1996), leaving a shortfall of SFr 72,000. If the total operating costs for SRI are converted to sterling at a rate of 2 SFr to £1 they come to just under £15 million. Compared with the 1996 budget of the BBC World Service of £137 million (not including the separate capital budget for new transmitters), I believe the Swiss government has one of the most efficient world service broadcasters. It also supports the Swiss economy in having all but one of its transmitters supplied by the Asea Brown Boveri works in Switzerland.

Table 11.3 *Milestones in the history of Swiss Radio International (SRI)*

1934	Swiss start up a national broadcasting service on MW (1st June) and a SW service from Prangins (1st August)
1935	Regular monthly SW broadcasts start in German, French and Italian
1939	Schwarzenburg transmitter site becomes operational (1st July) but is destroyed by fire one week later
1940	Rebuilt Schwarzenburg site resumes regular broadcasts in the autumn
1941	SRI begins regular broadcasts in English, Spanish and Portuguese and becomes a much listened to broadcaster during World War II
1946	First of three new high power 100 kW SW transmitter becomes operational
1947	SRI is allocated 71 frequency hours per day by the ITU
1954	SRI is given an additional mission; to strengthen links with Swiss abroad
1956	SRI proposes an Arabic service
1963	SRI is allocated budget for Arabic and French for Africa and Middle East
1972	First 500 kW SW transmitter and rotatable curtain is commissioned in May
1973	According to Gallup poll in the US, the Swiss SW service is ranked as one of the five most credible international broadcasters
1978	The name Swiss Radio International (SRI) becomes official (5th November)
1987	SRI signs relay agreement with China – two Chinese transmitters for SRI in exchange for China using two Swiss transmitters
1992	SRI starts up its first satellite channel, ASTRA, for European listeners
1993	SRI develops new strategy for the 1990s, geared to audiences and markets and for the creation of a continuous-relay programme, digital production and the use of satellites for all continents
1993	SRI launches transmissions over satellite INTELSAT-K for North, Central and South America
1994	SRI launches first of four planned satellite programmes over EUTELSAT for French audiences followed by an English language programme on a second ASTRA channel
1995	SRI introduces its new German and Italian satellite programmes on the first ASTRA channel
1996	SRI announces that it intends to broadcast to all continents by satellite. (SW to continue to be main gateway to non-European areas, and locally intensified and retained for European audiences until mobile reception by satellite becomes possible)

Chapter 12
Radio Nederland Wereldomroep

The Netherlands was not only the first nation to inaugurate radio broad-casting in Europe, but it was also the first European nation to start up a SW service. For centuries the Dutch travelled and traded abroad, and built up an empire in the Far East in the process. Like the British the Dutch saw radio as a means of communication with their far-flung colonies. It was not surprising then that the Dutch pioneered SW communications with their colonies, the Dutch East Indies (now Indonesia).

This particular SW route was not one of the easiest to conquer. The great distance required multi-hop ionospheric propagation over land, and in consequence the signals suffered severe attenuation. Transmission technology was the driving force behind SW to the East Indies [9], and the Philips Radio Company, which had begun by making lamp bulbs in the 19th century, consequently turned its attention to constructing the most powerful transmitting tubes in the world, to power the huge radio transmitters required.

After a series of extended SW trials in 1925 and 1926, in 1927 Philips inaugurated the world's first long distance radio telephone and radio broad-casting service from its SW transmitting station in Eindhoven, to the Dutch East Indies. (The first SW station I picked up in 1934 was PCJJ; the QSL card had an orange background and gave details of the transmitter output power and valve types.)

Of course the Philips company did not just make powerful radio transmitters, it also produced electrical goods. The logic behind its radio broad-casting was that it would generate a market for domestic radio sets, which indeed it did. Philips was the pioneering radio company in Europe, and the contemporary of Westinghouse in Pittsburgh, USA. The long-distance SW service to Batavia and other cities in the Dutch East Indies ran for more than ten years, ending in May 1940 when the Netherlands were occupied by Germany. The Philips factory at Eindhoven and its SW transmitting station were two highly strategic acquisitions, and the Germans were not slow to start using the SW circuit to Asia for themselves.

Figure 12.1   *Radio Nederland Wereldomroep (RNW)*

Meanwhile in London, the Dutch government-in-exile was able to get air-time on the transmitters of the British Broadcasting Corporation. Henk van den Broek, the former Paris correspondent of the Dutch *Telegraaf* newspaper, became the head of this Dutch Radio service. In September 1944, immediately after the liberation of the city of Eindhoven by allied armies, van den Broek flew to Eindhoven to head up a new radio broadcasting project. In conjunction with a small number of Philips employees, a freedom radio station went on air on 3rd October 1944 with the identification 'Radio Herrijzend Nederland', its broadcasts were aimed at the northern part of the Netherlands which was still under Nazi occupation. After the full liberation of the Netherlands van den Broek was asked to create a proper external radio service in several languages. All this took shape in 1946 and on 15th April 1947 the Radio Netherlands Foundation was officially inaugurated.

Radio Nederland Wereldomroep (RNW) today broadcasts in five languages (Dutch, English, Spanish, Indonesian and Papiamento) over its own facilities. Additionally the Projects Department supplies stations in Europe, Latin America and Africa with programming via satellite, in Portuguese and French. In addition, special television projects have included production in other languages such as Chinese. The studio facilities centre remains in Hilversum, where it has always been.

Indeed, one of my earliest recollections, at the age of 6 or 7, was watching my father tune into Hilversum on the dial of the family radio set and hearing the station announcement 'Here is Hilversum Holland.' Today, Hilversum studios and headquarters employs around 250 permanent staff and countless freelance contributors. Not all are employed in radio production because Radio Nederland Wereldomroep is also engaged in other activities such as international television in which it co-operates with Cable News Network (CNN), Atlanta, USA. RNTV supplies a weekly news contribution to the 'CNN World Report', a worldwide news exchange programme in which more than 120 countries participate.

RNW radio production: direct transmissions

Around 100 permanent staff are involved in programme production for direct transmission. This team is backed up by another group of freelance contributors, making up a closely knit team of presenters, newsreaders, editors, reporters, translators and producers. The Radio Netherlands newsroom serves all the regional departments by providing news bulletins on the hour and half hour. It also produces feature material written by specialists in the fields of European and international current affairs, economics and finance, culture, science and sports.

Programme transmission strategy is geared to reaching world audiences at prime time, so while radio transmission goes on around the clock, the programme times are arranged around the different time zones in the world. So, for example, the first transmissions in the early hours after midnight are directed to the Pacific Rim.

Over the years RNW has moved to using a number of delivery systems other than the SW bands, and transmissions are now available by four different methods: SW, MW, cable and satellite.

The SW service continues to serve all four corners of the world from SW facilities in Flevoland in the Netherlands, Bonaire in the Netherland Antilles, Madagascar and from leased facilities in Russia. Listeners in Europe are catered for by late evening 1440 kHz MW transmissions while those in Canada are reached by overnight MW and FM relays over the CBC network. RNW programmes are transmitted over Astra 1B in Europe and the North American Galaxy 5. The placement shift to using FM relay and satellite transmission is in line with current trends; although SW is still the main gateway for international broadcasting, whilst it is an excellent medium for talk radio it is inferior to FM for music. This is why there is so much interest in developing delivery systems which have all the advantages of classical SW broadcasting but with the ability to handle near-CD quality sound.

Transmission sites

The main SW transmitter site for Radio Nederland Wereldomroep is at Flevo-Zeewolde in the Netherlands. It has one 100 kW AEG Telefunken S4001 and four AEG Telefunken S4005 500 kW transmitters which went into service in 1982. At that time the latter transmitter was the most advanced in the world, and RNW was the first international broadcaster after Deutsche Welle to receive it. Flevo has 17 SW antennas of the directive certain type, and two non-directive antennas used for European coverage. The transmitter site at Flevo is used primarily for serving Europe, Africa, parts of Asia and the Middle East.

The second largest SW transmitting station of RNW is on the island of Bonaire in the Netherlands Antilles. Constructed in 1969 it has one 250 kW SW transmitter and two of 300 kW output power. The site has 22 antennas, all but one being directional, targeted towards North America, West Africa, Australia and New Zealand. This station employs a staff of about 45.

Completing the number of SW transmitting stations owned and operated by RNW is a second SW relay station in Madagascar. Similar to the Bonaire station, Madagascar has two 300 kW SW transmitters and 18 directional antennas for serving Africa and Western Australia. The Madagascar relay station also has a staff of about 45.

Figure 12.2   *RNW SW transmitter station at Bonaire, Netherlands Antilles*

Relay exchange and leasing agreements

In common with most international broadcasters RNW has relay exchange agreements. These are with Deutsche Welle, which carries programmes for RNW over its SW facilities at Julich and Nauen, in Germany. RNW also leases air-time on some ex-jammer transmitters located in the 'Commonwealth of Independent States (CIS)'. These SW transmitter sites are at Alma Ata, Dushanbe, Irkutsk and Petropavlovsk Kamchatskiy, (the latter at the easternmost tip of the former Soviet Union). According to RNW, these high power transmitters in the CIS enable RNW to increase its signal audibility in the Far East and SE Asia.

RNW: the voice of the Netherlands

One could describe the Netherlands as a nation that has reached political maturity in the world. Once the hub of one of the most prized empires, which stretched from the West Indies to the Dutch East Indies, the Netherlands of today seems to have few pretensions about being a world power, which is why it does not spend as much on its international broadcasting arm as the United States, Britain or Germany. It has more in common with the trusted international broadcasters of Switzerland, Sweden and Norway. In a world of political division it is good to hear balanced voices which observe tolerance and neutrality. The Netherlands, Norway, Sweden and Switzerland did not take part in the Western broadcasts to the Soviet union and Eastern bloc countries during the cold war.

The driving force at Hilversum is youth and enthusiasm. In much the same way that Amsterdam is a magnet for the youth of Europe, so the voice of Radio Netherlands International projects a warmth and enthusiasm over the airwaves and it exploits two of the Netherlands' most treasured beliefs: freedom of expression and independence of thought. Whilst some broadcasters are teaching people what to think, the Dutch believe in teaching people how to think.

RNW has earned acclaim from others for its originality in programming. During the winter months when many thousands of retired Dutch people go to Spain and southern France to escape the wet winter of the Netherlands, programmes are broadcast for this group via SW, satellite and over FM radio stations such as Radio Benidorm. During the summer months when up to 5 million Dutch leave the country on holiday, RNW keeps them in touch with a service providing news, weather and coverage of major sports such as the World Cup and the Tour de France (cycle race). The popularity of this new use for an international broadcast service is reflected in an additional listening audience figure of 1.7 million.

As an international broadcaster RNW does not have anything like the world audience of some, but at the same time it does not spend as much as the BBC World Service, for example, on its SW and MW transmission infrastructure, or on its language programming. Yet, the fact is that the Netherlands, along with France and Germany, is in the vanguard of trans-European broadcasting. RNW was one of the prime movers in the Archimedes System, a European Space Agency funded study group for a future digital audio broadcasting system, using three medium altitude satellites in highly elliptical orbits (HEO).

Chapter 13
The former Soviet Union

By whatever standards you care to measure, the old Soviet Union was truly immense. Its population was about 287 million, occupying a contiguous landmass extending from the shores of the Baltic Sea in Europe to the frozen wastes of Siberia, through to Far East Asia, past the Sea of Japan and round to the furthest tip of land bordering close to the state of Alaska. It encompassed no less than eleven time zones.

The products of a country, no less than its infrastructure, tend to reflect the nature and character of its people, a fact that partly explains why the Soviets built the biggest vessels, transport aircraft, helicopters, submarines, space launch rockets (exemplified by the Energiya SL-17 with a launch mass weight of 2690 tonnes) and the Mir Space Station which it launched in 1986. So it is not surprising that they also built the biggest and the most powerful SW broadcast transmitters and the largest SW transmission sites in the world.

The thoughts of Soviet scientists can be said to have run on different planes to those of their Western counterparts; their ambitions knew no limits, their horizons had no warp. As a consequence the word 'small' is an adjective that sometimes seems to have no place in the Russian language in the context of their technological achievements! Such things as miniaturisation and micro-miniaturisation were technologies they were happy to see other nations exploit. However, the ability of Soviet scientists to achieve was accompanied by a political desire to conceal, or play down the full extent of their achievements. For example, the world's first satellite in orbit, Sputnik 1, launched by the Soviets weighed 76 lbs (34 kg), but we now know that they had the rocket capability to place a much heavier mass into space. They chose not to do so, to retain secrecy on rocket development and to avoid the possibility of a violent reaction from the USA at a particularly sensitive period of the cold war. During the cold war I sought permission to inspect a state-of-the-art SW transmitter site. I was subsequently invited to see a site with 250 and 500 kW SW transmitters and rhombic antennas. What I was not told was that elsewhere the Soviets operated superpower transmitters with 1000 kW carrier power and massive, high gain curtain arrays with electronic slewing capability. At that time the high power SW transmitter was *the* weapon of the cold war, so it was sensitive to both sides.

We now know for certain that during the cold war Soviet expertise in the technology of high power transmitters and superpower transmission systems was ahead of the West. To dispel any ideas that Soviet-built transmitters are huge in size and therefore inefficient in performance, one must discard any preconceived notions of what constitutes good design. In the West we have come to accept the modern, highly compact 500 kW SW transmitter with micro-processor controlled auto-tuning as the ultimate in good design. Yet, all these innovative features were borne partly of necessity. It was the oil crisis of the 1970s that acted as the spur to creating energy-saving designs such as PDM (pulse duration modulation) and PSM (pulse step modulation). Small and compact transmitters kept down the cost of the transmitting station buildings, while fast automatic tuning features kept down the need for a large number of transmitters in a transmitter station.

By contrast, the USSR did not have an energy crisis, and with its huge landmass and central planning it certainly did not have the need to conserve space in its huge transmitter sites. Nor was there ever a pressing need to use minimum manpower at these installations – the Marxist policy saw to that. As a consequence of having few constraints the Soviets were able to fulfil their ambitions using the same philosophies that had enabled them to build giant space rockets. In the field of radio broadcasting in the AM spectrum, Soviet engineers concentrated on the most important feature – transmitter output power. They honed techniques for paralleling of high power transmitters to realise greater output power, and at many of the stations some transmitters were grouped by a twin parallel arrangement. Soviet scientists also developed and constructed directive antennas for MW broadcasting that were very many kilometres in length and with a forward gain in excess of 25 dBi. For SW broadcasting the most common antenna in use was the directive curtain array with slewing facilities.

It was this kind of transmission technology, dedicated to the goal of achieving high signal audibility, that enabled the Soviets to dominate across the spectrum, long, medium and short wave, to such an extent during the cold war. Whatever part of the dial your receiver was tuned to – and particularly on SW – there was the voice of Radio Moscow, with a clarity and strength rarely equalled by the BBC or VOA. The Soviets seemed to know the best choice of frequency for a particular target zone and were adept in other areas, like flexibility in power output. For example, a SW station with five 500 kW SW transmitters could swiftly be switched to operate as one 500 kW and two 1000 kW when ionospheric conditions began to deteriorate.

The transmission infrastructure of the old Soviet Union was no less impressive. It comprised some 300 transmitting stations of the high power

Figure 13.1 *View of 2000 kW 'Priliv' MW/LW transmitter built by RIPR, St Petersburg*

variety (100–2000 kW carrier power) and a much larger number of low power stations and radio jamming installations. The sheer size of the landmass, coupled with the fact that the USSR was made up of fifteen different republics, makes it virtually impossible for anyone to know accurately the true extent of this network, or the total megawatts of transmitter power that

Figure 13.2 *Output stage of 2000 kW 'Priliv' transmitter, with three Svetlana TY88N vapour-cooled tubes*

could be made available for foreign service broadcasting. Today the Soviet Union is a thing of the past, some of the former fifteen republics have become separate countries in their own right, and the remainder constitute the Commonwealth of Independent States (CIS). Dominating the CIS is the former Russian Soviet Federal Socialist Republic, now called Russia.

Although some of the member states of the CIS now operate their own foreign broadcasting service, by virtue of the fact that they possess the transmission sites once operated by Radio Moscow, it is Russia itself that still owns most of the former transmission infrastructure.

Radio Moscow International: transmitter network

Within the Moscow region, there were no less than eight high power transmitter sites, with an estimated total of over 80 SW transmitters, with carrier powers from 50 to 1000 kW. Outside the Moscow region and in geographical order from west to east were another twenty or so high power sites. These included sites at Sovetsk, St. Petersburg, Kovylkino, Samara, Ufa, Krasnodar, Volgograd, Yekaterinburg, Novosibirsk, Omsk, Angarsk, Irkutsk, Chita, Yakutsk, Khabarovsk, Blagoveshchensk, Komsomolsk, Ussuriysk, Vladivostok, Magadan and Petropavlovsk-Kamchatskiy at the easternmost tip of Russia.

As is the case with most countries, the transmitter sites are owned and operated by a separate authority, in this case the Russian Ministry of Communications, and since the end of communism Radio Moscow now has to pay rental charges to that authority. However, this leaves Radio Moscow free to negotiate for, and lease back, some of the high power transmission sites it used elsewhere during the cold war. It has concluded agreements with some of the former Soviet republics, namely Belarus, Moldovia, Ukraine, Armenia, Azerbaijan, Kazakhstan, Tajikihstan, Turkmenistan, Uzbekistan and Kyrgyztan.

If account is taken of the secret high power jamming transmitter installations that were spread the length and depth of the old Soviet Union, many of which have now been either dismantled or converted to broadcast transmitters, the overall total output of the transmitter installations was nothing less than staggering. Some of this transmitter network has subsequently been leased by international broadcasters in the West. Deutsche Welle, for instance, has taken a lease on two 1000 kW SW transmitters at Novosibirsk, which makes it the most powerful site in the world.

These relay agreements have proven to be of immense strategic value to western broadcasters, not merely because they are some of the most powerful installations in the world, but because of geopolitical factors; their location and proximity to countries such as China, North Korea and Laos means that VOA, the BBC-WS and DW can target these countries with high audibility broadcasts.

Figure 13.3   *RIPR 50 kW VHF-TV transmitter under factory test*

Table 13.1   *High power and superpower transmitters located in the former USSR now leased out to the BBC World Service, Deutsche Welle and Voice of America. (Some sites previously used for jamming.)*

Site	User	Type	Quantity	Carrier power (kW)
Novosibirsk	DW	SW	2	1000
Novosibirsk	VOA	SW	2	100
Samara	DW	SW	1	250
Samara	DW	SW	1	200
Irkutsk	DW	SW	1	250
Irkutsk	VOA	SW	1	500
Irkutsk	BBC	SW	1	500
Tashkent	BBC	SW	1	500
Tashkent	BBC	SW	1	250
Chita	BBC	SW	1	500
Chita	BBC	SW	1	500
Yekaterinburg	BBC	SW	1	250
Novosibirsk	VOA	SW	1	500
Krasnodar	VOA	SW	1	500
Petropavlovsk-Kamchatskiy	VOA	SW	1	500
Ussuriysk	VOA	MW	1	1000
Dushanbe	VOA	MW	1	1000
Tajikhstan	VOA	MW	1	1000
Kamo, Armenia	VOA	MW	1	1000

In addition, the BBC, DW and VOA lease air-time on some lower powered MW transmitters and on some VHF-FM transmitters.

Under the terms of lease the sites will deliver broadcasts to specified target zones – western authorities will have no responsibility for the operation of these sites, transmitter performance or selection of frequencies. According to engineers at the BBC-WS the transmissions from these Soviet-built transmitters produce consistently good signals in the target zones.

Siberia

Siberia means 'sleeping lands'. It is a land of frozen wastes where on a winter day there can be ambient temperatures of minus 37° C and there are regions of permafrost, which even during the summer can extend to two metres below the surface. It is a vast territory covering an area of more than four million square miles which extends almost half way round the world. To the north it borders the Atlantic Ocean, to the east the Bering Strait, the Sea of Okhotsk and the Pacific Ocean, whilst to the west it is divided from European Russia by the Ural Mountains and the Ural River.

Siberia was the place where the leaders in the Kremlin banished both criminals and dissidents to a life of misery and hardship. Because of their sheer isolation and remoteness – many of the towns can only be reached by aircraft – these frozen wastelands had other uses, they were the ideal place to locate high technology research facilities away from the prying eyes of western agents. Many of these towns devoted to military projects had no names and were identified by a code letter or number.

The frozen wastes of Siberia were the place where the Soviets constructed their most powerful SW and MW broadcasting installations, along with equally powerful jamming facilities. These ranged in carrier power up to 1000 kW. A jammer is synonymous with a broadcasting transmitter, the main difference being in the audio modulated signal. Technical requirements for the siting of 'sky-wave' jammers, that is, radio jamming carried out by target broadcasting via the ionosphere, apply equally to target broadcasting facilities, in that the transmitter facility should be positioned in a flat area with good electrical conductivity, at an optimum distance to project strong signals into the target zones.

High power SW transmitter facilities were constructed at sites such as Angarsk, Irkutsk, Krasnoyarsk, Omsk, Novosibirsk and as far east as Petropavlovsk-Kamchatskiy, on longitude 160 degrees E.

Novosibirsk, in Time Zone 7 hours East, is one of the many high power sites within Siberia. This facility served a dual role as a propaganda broadcast station and as a sky-wave jammer. Because it is within a one or two-hop path, via the ionosphere, to the cities of Western Russia, it was ideal for blanket jamming in Leningrad, Moscow and across the Baltic states.

Novosibirsk is equipped with more than thirty high power transmitters. It has 120 curtain arrays along with other antennas. In all, the antenna site covers more than two thousand acres – the largest in the world. To compare, it is more than five times larger than the next biggest SW station, at Gloria in Portugal (which was the hub of the SW network operated by Radio Free Europe/Radio Liberty during the cold war).

Altogether Siberia is thought to have had two hundred SW broadcast transmitter or jammers in operation up to the end of 1989.

Radio Moscow International: 'the Voice of Russia'

Radio Moscow first came on-air in September 1922 with transmitter station RV-1, broadcasting to the peoples of Moscow and surrounding regions. In 1925 a second broadcasting centre came on-air in Leningrad and these two cities became the cradle for radio broadcasting within the USSR. The pace of development in this new branch of science was rapid and by 1926 the race for higher power was on. From the early years the Soviets used all three AM wavebands, long, medium and short wave, for its home broadcasting network, gaining an early introduction to the merits of the short waves for spanning long distances.

In Russia there is no doubt as to who was the father of radio. Professor Popov, a Russian scientist, was experimenting with wireless transmission before the young Guglielmo Marconi had arrived on the scene in England

Figure 13.4 *Novosibirsk SW station, Siberia*

in 1896. The Soviet naval fleet had wireless in 1914. The world's first public service broadcast was from the Russian cruiser Aurora, moored on the Neva River within sight of the Winter Palace in October 1917. That particular broadcast signalled the start of the Russian revolution.

Russian foreign service broadcasting got under way in 1929 with Radio Moscow calling the world on SW and by the end of that year it was broadcasting regular programmes to Europe, North and South America, Japan and the Middle East in English, French and German. The Soviets were quick to perceive the importance of broadcasting in other languages, something which did not start in Britain for nearly another decade. From 1935 onwards Radio Moscow devoted much time and effort to warning the rest of Europe about the dangers of German national socialism.

By 1938 propaganda broadcasting in Europe had become a fact of life, with the four big players, Britain, Germany, Italy and the USSR, representing British imperialism, German national socialism, fascism and communism respectively. The emerging medium of international broadcasting was serving notice of its future potential as an instrument of war. By 1941 the USSR was broadcasting to the world in 21 languages and a decade later the output from the studios of Radio Moscow reached an all time high of 2094 programme hours per week, broadcast in 80 different languages. This figure was by then greater than the output of VOA, although slightly less than VOA and RFE/RL combined.

Although programme hours broadcast is used by monitoring agencies, such as the BBC at Caversham Park in its league tables for the top 30 international SW broadcasters, the information does not present a fully accurate picture because it fails to take account of transmitter output power. In this area the Soviets certainly dominated the rest of the world's international broadcasters. The USIA and VOA have never deviated from their admission that here they have never caught up with the USSR.

One of the reasons for Radio Moscow's domination of the international HF spectrum is that radio broadcasting has always played a powerful role in its domestic broadcasting, exploiting the AM wavebands and especially SW because of its suitability to cover the gigantic landmass of the Soviet Union. With such a dependency on SW for home broadcasting it was natural that the Soviets excelled in developing high power transmitters for international broadcasting.

Ironically it was the Soviet's dependence, to a large extent, on SW for national radio that made them vulnerable to incoming broadcasts from RFE/RL, the BBC-WS and VOA. SW listening in the USSR was not a hobby as it was in the West, but more of a necessity. The ordinary citizen in Russia was likely to possess a SW receiver, with which they could also receive Western broadcasts, and this ability played a pivotal role in bringing about 'perestroika', and the subsequent collapse of Soviet-style communism in the USSR.

Radio Moscow had a wide and stable listening audience in every continent and region of the world, consisting of several different groups, such as those who admired and respected the Soviets for their achievements and some from the academic world in the West. Developing countries whose economies had been exploited by capitalism, or who had suffered in other respects from imperialism, also provided significant listening audiences. Combined, these enabled Radio Moscow to build up a listening audience greater than that of VOA. The end of the cold war and the collapse of communism in the USSR and elsewhere could have been catastrophic for its foreign service broadcasting, and the fact that this did not happen speaks volumes for the loyalty of listeners to Radio Moscow.

Radio Moscow did experience significant change with the ending of the cold war, which removed its *raison d'être*. In this respect Radio Moscow was not all that different from its Western counterparts, whose build up over the cold war years also owed to ideologically committed governments with a common dedication – the overthrow of communism in the Soviet states. However, whereas the Western powers restructured international broadcasting to suit new policies, Radio Moscow found itself with neither the financial resources or the entrepreneurial skills to do the same.

Shortly after the end of the cold war Radio Moscow lost its status and became part of the large domestic broadcaster Ostanko. In 1993 the Russian government seemed to go some way to reversing this decision when President Yeltsin signed a decree which took a step toward restoring Radio Moscow to its former status. Since then, however, progress in that direction has been slow. Russian foreign service broadcasting has also experienced difficulties of another kind, related to programme production. Before the collapse of communism, Radio Moscow's broadcast policy was clear; it simply had to follow the bidding of the Program Directorate, and no programme was broadcast without it first being vetted by that Directorate. When perestroika came and controls were loosened, the foreign broadcasting service had to begin to formulate its own policy, a task not made any easier by the fact that the production staff had no previous experience in marketing or working in a free media.

An ongoing problem facing Radio Moscow is finding adequate funding. The lifeblood of a foreign service broadcaster is its network of offices, news reporters and stringers, and running such a network calls for expenditure in hard currency, a commodity which is presently in short supply in Russia. As a result the service is losing some of its best reporters with the consequence that it has had to cancel some of its foreign programmes. The likely outcome of this is that Radio Moscow will increasingly direct more of its broadcasts to the other former republics of the old Soviet Union. The danger of such a move will be that Radio Moscow will in effect be competing with the national broadcasters for the same listening audiences.

It has also experienced an identity problem. For more than sixty years it was Radio Moscow, but after perestroika it adopted the name Radio

Moscow International (RMI). Then it began calling itself World Service and has now settled on Radio Moscow International 'Voice of Russia'.

Notwithstanding all the difficulties that Russia's foreign service broadcaster has experienced, and given the fact that much of its transmission structure is located in territories of other former republics, RMI has managed to date to retain the largest and the most powerful transmission infrastructure in the world. Although Radio Moscow is not as powerful as it was during the cold war years it remains a powerful voice on the airwaves. Figures supplied by the BBC International Broadcasting Audience Research unit gave a weekly output of 1300 programme hours in 1996, giving it third ranking in that league table, behind the USA and China. Radio Moscow has access to more than 200 SW and MW transmitters. More than 20 of these transmitters have output power between 1000 and 2000 kW carrier power.

Today, Radio Moscow is competing with other international broadcasters for the same listening audiences across a range of 48 languages. And just like Voice of America, the BBC World Service and Deutsche Welle, it has to ride the fine line between editorial independence and broadcasting in the interests of the state. The head of the English service department has said that RMI 'will always reflect the viewpoint of the Russian Government, but this should not prevent it from stating other viewpoints.

Chapter 14
The Balkan region

Through the period of the cold war and up to 1990 the Socialist Federation Republic of Yugoslavia (SFR) consisted of six socialist republics: Croatia, Macedonia, Bosnia-Hercegovina, Montenegro, Slovenia, and Serbia. The last of these had two separate districts – Vojvodine bordering Hungary, to the north, and Kosovo bordering Albania to the south.

For more than forty years SFR Yugoslavia was an important satellite of the Soviet Empire and an important member of the Warsaw Pact nations. In most respects the SFR was different to Eastern bloc countries such as Czechoslovakia or Hungary, not least in the degree of autonomy which it enjoyed. Partly because of its remoteness from Moscow, and partly because of the political importance of these Balkan states, but chiefly due to the power exerted by its head, Marshall Tito (who was much respected by the Kremlin), the SFR came to enjoy an autonomy denied to other Warsaw Pact member states and it traded with the US and Western Europe throughout the cold war.

The Balkan states represented a unique federation of different ethnic groups with differing shades of Christianity and Islamic beliefs. Nor were the differences restricted to religion; there were also substantial political as well as economic variations. In World War II, for instance, whilst Serbia was on the side of Russia (and therefore the Allies), Croatian leaders were doing deals with the Nazis. At the same time the Allies were working with the Kosovars and the Albanians, a fact little remembered during the 1999 conflict between NATO and Serbia. Presiding over this hotbed of Balkan states was Tito. During his life he held the Balkan states together to the extent of ensuring that the ethnic minorities had rights and political status equal to the majority in the more important republics like Croatia, Serbia and Bosnia. Tito died in 1980 before all his aims and ideals had been realised. Tito was half Croatian, half Serbian and was replaced by his disciple Slobodan Milosevic, an ardent communist possessed with many of the same ideals and beliefs as Tito. Nine years after Milosevic came to power the cold war ended and the Federation of Yugoslavia lost its political edge. Aid from Moscow and from the West began to dry up and political tension began to ferment, partly as a result of cold war propaganda from the West urging the Eastern bloc countries to fight for nationalism.

Croatia and Slovenia were the first to seek national independence and to break away from the SFR, prompting others to wish for the same. Croatia and Slovenia seceded from the Yugoslavia federation and became right wing republics. The damage done to the federation by these events was catastrophic. For example, Croatia now possessed almost the entire coastline of the Adriatic and with it the marine and shipping industries along with Rjeika's deep sea port, the biggest in the Adriatic. Croat disputes with Serbia extended into Bosnia, where there lived a million Croats along with a lesser number of Serbs. Thus the seeds were set for the first 'ethnic cleansing' war in Bosnia-Hercegovina, with Croats and Serbs fighting for possession of territory. While all this was taking place Milosevic was having problems in Kosovo where Albanians outnumbered Serbs by a ratio of 9 to 1 and had declared Kosovo to be a separate republic. Thus, over a period of a few years, Serbia – once the very heart of the Yugoslav federation – had lost much of that federation and was now in danger of losing districts of greater Serbia itself. To the Serbs Kosovo is not merely a district or province of Serbia. It is arguably the spiritual heart of Serbia, a fact overlooked by many Western nations, during the NATO conflict with Serbia. Kosovo is where Serbia began, the historic heartland and the symbol of resistance. On 28th June 1389 the Serbian Prince Lazar died in the Battle of Kosovo attempting to resist a Turkish invasion. Lazar's remains are still preserved and Kosovo is where great monasteries were built to keep alive the spirit in (for the Serbs) virtually a holy land. At the same time, over many years Albanians have migrated into Kosovo in such numbers as to regard it as their home too.

Given this history, the background to the 1999 conflict between NATO and Serbia is far from simple. Although there is widespread acceptance of the claim that Milosevic is guilty of major 'ethnic cleansing' atrocities in Kosovo, during the Bosnian war the Croats were engaged in similar activities in Bosnia, as well as in the 1994 war with Serbia. One could also speculate that as Milosevic is one of the few remaining communist survivors from the cold war, his vilification by the West is to be expected, but this may be too simplistic an analysis. Serbia is the last remaining piece of the communist empire which, as long as Milosevic remains leader, could be seen to constitute a threat to the now right-wing states which border it, specifically Croatia, Slovenia, Hungary, Romania and Bulgaria.

Radio broadcasting using very high power AM has always played an important role in the media in the former Yugoslav republics, as it did in the former USSR and Eastern bloc countries. The six republics of the former SFR totalled about 26 million people, yet the federation had a total of more than 17 high power transmitters, mostly MW, with carrier powers of up to 2000 kW (see Table 14.1). Serbia, for instance, had one 2000 kW

Table 14.1 *Locations of SFR high power AM transmitters (100–2000 kW), 1986*

Republic	Location	Transmitter power (kW)
Serbia	Beograd	2000
	Beograd	200
	Aleksinc	200
Slovenia	Ljubliana	600
Croatia	Tovanik	300
	Zadar	1200
	Deanovec	100
	Biograd	1200
Montenegro	Titograd	100
Kosovo	Pristina 1	1000
	Pristina 2	100
Vojvodina	Novosad 1	150
	Novosad 2	600
Macedonia	Skopjie 1	1000
	Skopjie 2	100
Bosnia-Hercegovina	Sarajevo 1	600
	Sarajevo 2	100

Note: Total AM power in 100–2000 kW range is 8850 kW. This figure excludes the large numbers of lower powered AM and FM stations

device and a further four of more than 1000 kW power. These were either of Russian build or were supplied from the SFR state-owned manufacturer RIZ of Zagreb. Since the break-up of the federation Serbia has added a further 1000 kW MW transmitter in Kosovo for foreign service broadcasting. At the time of writing it is assumed that this and other Serbian radio broadcasting infrastructure has been a target for NATO bombing. This recent conflict, NATO's first and the first time it has embarked on a war with a former ally, has once again brought radio and television into focus; first as a propaganda weapon used by NATO and Serbia, and secondly as a target for NATO bombers.

In fact only the techniques used in war have changed over time – the media have always been a target. In World War II, although radio propaganda played a powerful role, neither side deliberately targeted broadcasting installations, but rather preferred to capture these intact and turn them round to broadcast to the enemy. Europe's only superpower broadcasting station in Luxembourg was a classic case; it changed hands twice in 1940 and then again in September 1944, without so much as breaking a tube or gramophone record. The Gulf War saw a change in objectives, as the Allies inflicted massive damage to Iraq's radio broadcasting installations, to the point of total destruction.

The Serbo–Croat war of 1991 saw a different technique. Serb aircraft conducted precise low-level attacks to disable antennas but they did not attack studio centres in cities. This was a fairly humane approach because transmitters are mostly unattended and, in any case, it is always more effective to put transmitters out of action rather than studio programme centres. In this 1999 conflict we have seen the opposite technique. NATO appears to have sought to cause as much material damage as possible to the studios, presumably with the twofold objective of diminishing the ability of the Serb authorities to communicate with the people and also to demoralise civilians. One would have thought the attack on the main television studios in Belgrade, which killed production and programming staff, to have been unnecessary as it would have been more cost-effective to have made a single air strike at the TV transmitter sites.

At the time of writing, it is not known how much remains of the Serbian AM infrastructure after the NATO conflict. It is also assumed that the Serbian foreign service on SW has been disrupted.

Croatia

Croatia, with a population approaching five million, mostly of Croat origin, achieved independence from the SFR in 1992. It has a culture that extends back over more than 900 years and is a country which straddles Western and Eastern Europe, with a coastline onto the Mediterranean.

Zagreb, the capital of Croatia, was the birthplace of Balkan broadcasting in 1925. Much later, when Croatia became a part of the Socialist Federation Republic of Yugoslavia (SFR), Zagreb was the hub of Yugoslav radio and television broadcasting, and the headquarters of the state-owned company Radio Industries Zagreb (RIZ), a major European manufacturer of telecommunications, radio and television equipment. Today, Zagreb is still the headquarters of the now privatised RIZ Transmitter Company. Hrvatska Radio and Television (HRT) is the authority responsible for all radio and TV broadcasting throughout Croatia with Hrvatska Radio (HR) responsible for radio broadcasting.

HR operates a comprehensive network of sixteen AM transmitters all on MW, the three most powerful being Deanovic (100 kW), Tovarnik (300 kW) and Zadar with a 1200 kW transmitter – one of the most powerful MW stations in Central Europe. HR also operates a substantial quantity of FM radio stations, mostly situated in cities and along the coastal region of the Adriatic. The 1200 kW transmitter at Zadar reaches out to audiences throughout central Europe, the Mediterranean, Italy and as far as western Europe. This is due to the combination of a number of unique features; its radiated power, its four mast directional radiator and its topographical location.

Foreign service broadcasting

When the fledgling state of Croatia formally achieved independence in
1992 it found itself without a voice in the international SW bands, the
reason being that the foreign service SW stations of the former Yugoslavia,
of which Croatia had been part, were located in other territories that now
formed Serbia and Bosnia-Hercegovina. As a consequence a SW service was
high on the list of priorities for the new state of Croatia, although it was low
on the list of fiscal budgets because of the many other demands upon the
new government.

Preliminary census studies carried out in 1994 showed a population
figure of some 4.78 million people living in Croatia. Further studies
reinforced the need for a foreign service due to the many descendents of the

Table 14.2 *Milestones in the history of Croatian radio broadcasting*

1925	Radio Zagreb goes on air, broadcasting on 350 m from St Mark's Square, Zagreb (15th May)
1927	Radio Zagreb begins foreign co-operation; transmits programme on SW through the station at Schenectady, USA
1928	Radio Zagreb becomes full member of the UIR as the representative of the Kingdom of the Serbs, Croats and Slovenes
1934	Radio Zagreb moves to better site on the banks of the River Sava
1940	Radio Zagreb shareholder's company is nationalised and receives substantial government funding from Yugoslavia
1942	Power of Radio Zagreb is increased to 10 kW, and regional stations are set up in Sarajevo and Banjaluka
1943	Radio Osijek goes on air
1945	Radio Split goes on air
1949	Radio Zagreb gets a high power MW transmitter of 135 kW
1956	On 13th anniversary the first TV transmitter comes into service, ready for coverage of the Zagreb World Fair
1968	Radio Zadar goes on air
1970	Tovarik goes on air with a 300 kW MW transmitter
1975	Radio-TV Zagreb begins colour television broadcasting
1976	50th anniversary of radio and 20th anniversary of TV. Programmes of Radio Zagreb (and regional stations) begin broadcasting programmes from 39 MW transmitters and 22 UHF-TV transmitters and repeaters. TV Zagreb second programme goes out over total of 75 TV transmitters and repeater stations (total output power 173 kW)
1986	Radio Zadar goes on air with 1200 kW, the most powerful MW station in Central Europe and the Balkans
1990	A transmit-receive satellite earth station with 10.6m dish is added
1990	Croatia declares itself to be an independent state and Radio-Television Zagreb is renamed Hrvatska Radio Television (HRT)

Source: *Croatian Radio and Television* May 1991 (on 65th anniversary), HRT

Croat peoples who had emigrated to North America. Accordingly, with the assistance of the RIZ Transmitter Company, a new 100 kW SW transmitter was acquired for the specific objective of providing a foreign service beamed to Europe and North America on an azimuth bearing of 305 degrees, located at the SW site of Deanovic. Notwithstanding the modest power of 100 kW due to budget restraint, good reports have been received from listeners in Europe and North America, including the BBC monitoring station at Caversham. All programmes are preceded by announcements 'This is Croatian Radio' and appropriately the station music is a tune to Dubrovnik's poem *Lovely Dear Sweet Liberty* played on the celeste. Transmissions go out on 5895, 7370, 11635 and 13830 kHz.

Slovenia

Slovenia is one of Europe's newest independent states. Once a province of SFR Yugoslavia, this now independent country, with a population of less than 2 million, owes its independent existence in part to the collapse of communism in Europe and the Soviet republics. The promotion of patriotism or nationalism as an ideal was one of the tools used by the West. The break-up of Yugoslavia, the Serbo–Croat war and the separation of Czechoslovakia into two separate republics were in part consequences of the encouragement from the West for regions to seek goals of nationalism.

The national language, formerly Serbo–Croat, is now the original Slovene but with German and Italian widely spoken. The radio and television broadcasting authority is RTV-SLO, the Slovene Radio and Television company. RTV-SLO is a public, non-profit radio broadcasting organisation, financed through subscription which accounts for some 65 per cent of its total income, with the balance derived from advertising.

For radio broadcasting the topography of the country – mainly alpine – means that AM is the best medium. Slovenia has ten AM stations dotted around the country, the highest powered, and the flagship of RTV-SLO, is Domzale on the outskirts of Ljubljana. Until 1993 Domzale had a 600 kW MW transmitter operating on the frequency of 918 kHz. This was then replaced with a new 300 kW MW transmitter, a DX300 supplied by the Harris company. Notwithstanding the chief dependency on AM, RTV-SLO has some 20 FM stations, catering for the towns and cities with higher population densities. RTV-SLO also places great importance on being able to reach Slovenians resident in other countries in Europe, including first generation expatriots working in Italy, Austria and elsewhere.

Table 14.3 *Milestones in the history of Slovenian broadcasting*

1924	Ing Marij Osana starts up experimental broadcasting with home-built radio transmitter
1928	First serious experimental broadcasts begin from Radio Ljubljana and regular service inaugurated on 28th October with 2.5 kW power
1939	Contract concluded stipulating further expansion to 20 kW power
1941	German Luftwaffe destroys Radio Ljubljana site at Domzale on 11th April, Radio Ljubljana taken over by the Italians
1941	Radio Krica starts to transmit illegally and changes secret locations 23 times without being discovered (operated by the Liberation Front)
1944	Liberation Front now transmitting from liberated territory at Bela
1945	Radio Liberation very active broadcasting all day from Ravna Gora
1945	Yugoslavian People's Army captures Trieste on 1st May. Radio Free Ljubljana and Radio Free Maribor begin radio broadcasts
1951	New 135 kW MW transmitter is set up at Domzale
1955	Regular VHF-FM service starts up at Nanos, Kum and Pohorije
1958	Regular TV service from Radio Ljubljana as part of Yugoslav Radio and Television broadcasting service
1963	Fourteen new MW transmitters go into service
1973	New 600 kW MW transmitter enters service, 300 kW MW transmitter installed on Beli Kriz above Piran
1980	Construction of new transmitting centre at Krvavec begins. New microwave link between Ljubljana and Maribor brought into service
1991	Ten day War of Independence. Yugoslav aircraft attack and damage and destroy RTV transmitter stations at Kum, Nanos, Krvavec, the Ohorje, Domzale and Boc

Chapter 15
RFE/RL comes out of the cold

Radio Free Europe/Radio Liberty, which merged in 1974, is an international broadcaster with a unique history in more than one sense. Once the scourge of the Soviet Union and highly feared by its leaders in the Kremlin, this broadcasting agency achieved more than any other international broadcaster, or indeed any other instrument of war, to bring about the bloodless revolution in the Eastern bloc countries and the overthrow of communism in the USSR.

One of the tools used by RFE/RL was the programme content of its broadcasts. The other, which perhaps played the pivotal role, was the way it targeted its SW transmissions with high accuracy. For example, in the case of Czechoslovakia, it targeted the Czech and Slovak regions separately, thus recognising the political differences which existed between the rural Slovaks and the more urban Czechs. Such differences created more work in RFE/RL, requiring separate programming, but made it easier to achieve the end objective – the undermining of Soviet power.

Altogether, RFE/RL transmitted specially prepared programmes in more than 20 different languages to all the 'enslaved' countries and the different republics within the USSR. As a medium for spanning long distances SW broadcasting is peerless. It can span oceans, cross frontiers, its baggage cannot be censored and there is no third party involvement. It reaches out to the poor and the rich with equal facility bringing locally uncensored news and the listener's only investment is the ubquitious radio receiver with the facility for receiving SW.

SW transmission is one of the oldest forms of radio communication, although 60 years ago it was more art than science. Today there still exists that small element of art, but it is also a highly developed science which calls for an understanding of the vagaries of the ionosphere. It is in such areas of science that RFE/RL has emerged head and shoulders above the rest of the international broadcasters. RFE/RL was originally two separate broadcasters, both run by the Central Intelligence Agency (CIA). They were each allocated the same mission, though with different focuses: to bring the voice of freedom to the peoples of Eastern Europe and the Soviet Republics, with the aim of overthrowing communism. Achieving this goal called for adaptive, creative and innovative scientific skills in order to be able to deliver reliable and consistently strong signals to precise target zones in the presence of powerful Soviet jammer installations.

It is no overstatement to say that RFE and RL achieved more than was expected of them, and more than was thought possible by some in the political arena; but what RFE/RL also elevated was the science of SW transmission to new heights.

RFE and RL were instruments of the cold war that began in 1948 and lasted over forty years. No one thought this war would go on for forty years and when the end came it was sudden and unexpected, but most important of all it was nearly bloodless. It was a war fought with words, not bullets; it was paid for by American taxpayers, funnelled in the beginning via covert channels in order to distance the US government from the activities of the CIA. But the funding of RFE/RL, first by the CIA, and later the Board for US International Broadcasting, was described by Washington in December 1991 as one of the best investments ever made by the US government [1].

Both RFE and RL were classified by the US government as surrogate or covert broadcasters. The original name Radio Liberation was changed in the 1960s to Radio Liberty to make its intent less obvious. In fact the Soviets were under no illusions about the purpose of RFE/RL or its intent, or that it was a tool of the CIA. Kruschev on more than one occasion described the CIA as 'the most dangerous organisation on the face of the earth'.

Voice of America, in contrast to RFE/RL, was a legally registered US international broadcaster. It was overtly respectable and had a different mission. VOA was formed in 1942 to be the voice of the nation, a voice that would help to counter Nazi propaganda in World War II and later to disseminate American views to the whole world whilst RFE/RL concentrated on the Soviet Union.

There is nothing new about using subterfuge broadcasting agencies. First used by the British to cause unrest in the Third Reich, Germany and America also made use of illicit and covert radio stations during World War II, so the setting up of RFE and later RL was just a continuation of a well established practice. In the twilight world of surrogate and covert broadcasters it is a fairly safe assumption that any radio station with the words 'free', 'peace' or 'liberty' in its station announcement is a revolutionary broadcaster. Where RFE/RL has distinguished itself from covert propaganda broadcasting in World War II is that in the latter radio propaganda was used as an adjunct to conventional warfare. However, RFE/RL operations were clearly aimed at bringing about the overthrow of Soviet communism by non-violent methods. That RFE/RL succeeded in their mission to bring the voice of freedom to the peoples of such a huge landmass is a

Figure 15.1 *Advertisement in WRTH for Radio Free Europe/Radio Liberty*

tribute to the programmers, engineers and scientists who devised and constructed RFE/RL SW transmission facilities in Germany, Portugal and Spain.

Barrage broadcasting and jamming

Wars act as a spur to technology and the cold war was no exception. The Soviets honed techniques using sky-wave and ground-wave jamming installations to prevent Western broadcasts getting through to the Soviet people. The West for its part devised a number of techniques designed to maximise its signal penetration into the USSR. One of these measures was barrage broadcasting, where the same programme is broadcast over a large number of transmitters working on different frequencies, with the hope that some of these SW transmissions will get through the jamming net.

The USSR never denied it was jamming certain broadcasts from the West. Its stance on the subject was that a country is entitled to take whatever steps are deemed necessary to jam radio broadcasts intended to exploit a political situation, or to incite the people of a country to acts incompatible with law and order. Viewed in retrospect, the propaganda war where one side used barrage broadcasting whilst the other tried to jam it was an ascending war of oscillatory antagonism, and as safe as playing handball with live grenades! It could have easily escalated into a real war.

The role of the CIA

The success that RFE/RL achieved in getting its broadcasts through to the Soviet bloc was largely a result of the early work by the CIA. In the early years after World War II Germany was a paralysed and defeated nation and the CIA was in control of much of Germany. It made headquarters for RFE and RL in Munich, whilst at the same time acquiring abandoned airfields once used by the German Luftwaffe, which made ideal sites for SW transmitter facilities because they were on large flat plains and far from urbanisation, whilst also being strategically located for broadcasting to Russia and Eastern Europe on a single-hop basis.

A second key factor was the ability of the CIA to recruit a network of dissidents from the European Soviet-dominated countries like Poland and Czechoslovakia. These dissidents formed the nucleus of the radio programming staffs, were united in their dislike of communism but in other respects were diverse in character, and included liberals, academics, delanded nobles, royalists, reformed nazis and other dubious characters. These exiles were sought after by the West – some became celebrities and some even became rich. Being a dissident at that time became a profession as it brought a CIA salary. When the CIA first used some of these dissidents over the air it encouraged others behind the iron curtain to seek their fortune with the CIA and its broadcasting agency. This policy of the CIA payed huge dividends in that RFE and RL, with their carefully prepared programme material, began to build up a core of regular listeners who would hear first-hand information on matters such as internal corruption within the Soviet empire.

Finally, the CIA had its agents on the ground in the USSR and the Soviet bloc countries like Czechoslovakia, Hungary, Poland and Romania, with SW receivers, and these agents' duties included reporting back to Munich headquarters on such things as signal strength and effectiveness of the RFE/RL SW transmissions.

Soviet sky-wave jamming

The selection of Portugal and Spain as 'host' countries for the siting of SW transmitter facilities was in part determined by the politics of those countries, but also took into account their geographical location to give an advantage in overcoming the effects of Soviet jamming. RFE/RL described this phenomenon as 'twilight immunity'.

Jamming practised by the Soviets and satellite countries was highly sophisticated. It involved the use of very high powered sky-wave jammers, capable of covering large areas, augmented by many low powered jammers with antennas designed to maximise ground-wave coverage. These local jammers had a very limited range because of the rapid attenuation of ground-wave signals at the high frequencies, but they were nevertheless effective both day and night and in densely populated areas were a cost-effective method for interfering with RFE/RL transmissions.

Sky-wave jammers, on the other hand, together with RFE/RL transmissions, depended on ionospheric reflection to reach their target areas, this reflectivity being the result of solar illumination of the ionised layers. However, because RFE/RL SW sites in Portugal and Spain were lying far west of the target countries and the sky-wave jammer sites, there was a period during the late afternoon or early evening of each day when the reflection zone or mid-path point of the jamming signals had passed into darkness, with a corresponding loss of transmission. The path mid-point of the RFE/RL programme transmissions remained in daylight condition, however, thereby enabling these transmissions to reach their target areas for an additional few hours, or until solar illumination of the mid-point path had ceased.

Other techniques which RFE/RL employed to combat jamming included hints and tips to their listeners on how to devise directive antennas, and how to construct antennas which did not betray their presence to the authorities.

Technical monitoring

RFE/RL maintained three regular monitor stations in Berlin, Vienna, and Thessaloniki. These stations were equipped with a wide range of monitoring equipment, and checked the signal strength of desired and undesired (i.e. jamming) transmissions in the broadcast bands. They also checked RFE/RL transmissions on the hour continuously. This data was collated and analysed by RFE/RL propagation experts located in Munich and New York, who then estimated the effectiveness of RFE/RL transmissions and their relative performance against other broadcasters in the same broadcast bands.

The location of these monitor stations were chosen to be as near as possible to the target areas, so that the effectiveness of the Soviet sky-wave jamming installations could be measured, to enable RFE/RL to take any counter action necessary.

How the Western broadcasters performed during the cold war

The cold war lasted slightly more than forty years and during that time the programme hour weekly output from the main participants in the West rose. The propaganda output figures for the five main players: USA, Federal Republic of Germany, Britain, Israel and Canada, are shown in Table 15.1.

Figures for the USA include VOA, RFE and RL, with VOA output being slightly higher than the combined total of RFE and RL. For example, the 1991 figure for RFE/RL combined was 1038 programme hours. Also within the overall total of 2401 programme hours for the USA are included 14 hours for 'Radio Free Afghanistan' and 162 hours for 'Radio Marti'. In the case of VOA, the BBC-WS and Deutsche Welle, the output represents

Table 15.1 *Cold war propaganda output (programme hours per week)*

Country	Broadcaster	Decade					
		1950	1960	1970	1980	1990	1991
USA	VOA ⎫ RFE ⎬ RL ⎭	497	1495	1907	1901	2611	2401
FRG	Deutsche Welle	0	315	779	804	848	841
UK	BBC-WS	643	589	723	719	796	797
Israel	Kol Israel	0	91	158	210	253	244
Canada	RCI	85	70	76	72	96	96

Source: IBAR

the total of broadcasts directed around the world, whereas in the case of RFE and RL these broadcasters directed their total output to the USSR and the Soviet-dominated Eastern bloc countries.

According to highly placed sources, the order which these broadcasters contributed to the defeat of the Soviets in the cold war is as follows: in the first place by an overwhelming margin RFE/RL; in second place the BBC-WS; third Deutsche Welle; fourth VOA; then RCI and Kol Israel.

A brief history of RFE/RL

RFE and RL were originally two separate broadcasting agencies with related but different missions. The former was set up in 1949; its mission was to orchestrate resistance to Soviet doctrine within the Soviet bloc countries in Eastern Europe. Radio Liberation formed in 1952 and had an even more challenging task: 'to liberate the peoples of the USSR from the yoke of communism' (in the words of JF Dulles). Both these broadcasting agencies were products of an era when America was under the influence of Senator McCarthy and the brothers John Foster Dulles and Alan Welsh Dulles. When President Eisenhower made John F Dulles Secretary of State, and approved the appointment of his brother to be Head of the CIA, it was to have a far-reaching effect on US foreign policy. John F Dulles possessed a hatred of communism that amounted to a personal evangelical mission. In his own words 'The Iron Curtain would be rolled back, the peoples of the captive nations no longer abandoned to Godless terrorism. Unrelenting pressure would make the communist leaders impotent to continue their monstrous way' [12].

The idea of the CIA running two propaganda broadcasting agencies, RFE and RL, was just an extension of the power over which the CIA had control. It had virtually unlimited budgets, ran its network of spies, recruited its own assassins, ran a mercenary army and had its own fleet of aircraft under the control of Alan W Dulles, whilst John F Dulles as Secretary of State had indirect control of America's public service broadcasters NBC, CBS and ABC.

Germany at that time was not only a defeated nation, it was also occupied, and just the right place to kick-start the cold war and possibly World War III. The appointment of Eleanor Dulles (their sister) as the US State Department representative in Berlin brought even more power to the CIA and the Dulles brothers. Berlin got its own propaganda radio station, RIAS, and now Germany became the cockpit for the cold war.

The intentions behind the creation of RFE and RL were made blatantly obvious from the beginning. Radio Liberation, as it was first called, was under control of the American Committee for Liberation of the Peoples of Russia, before it was altered to be the American Committee for Liberation from Bolshevism and eventually the Radio Liberation Committee.

There was nothing half-hearted about the creation of RFE and RL. With the US Army in Germany and CIA men thick on the ground, not only in West Germany but in occupied countries in Eastern Europe, and even in Russia, the CIA was well placed to recruit a core of dissidents to run the radio programming under the control of Americans in Munich, Berlin and New York. The types of programmes broadcast were intended to create social and political unrest, and eventually bring about revolutions in the different countries and the Soviet republics. In short, both broadcasters were free to use whatever means necessary to achieve their objectives, on the basis (most probably) that success was self-legitimising.

By 1956 Radio Free Europe employed about 2000 people, including 600 German technicians, but the all-important programming staff consisted of about 450 Poles, Czechs, Slovaks, Hungarians, Romanians and Bulgarians. They were united in their hatred of communism but otherwise diverse in character. The Hungarian uprising in 1956 was one high point in the RFE success story. In early October 1956 the people seized control of an amateur radio station in Hungary and street fighting broke out. As the fighting intensified, RFE broadcast messages right up to 7th November. These were intended to make the people believe that American forces were ready to come to their aid. On 8th November, freedom radio stations appealed to Radio Free Europe to enlist Western aid just before the Russian tanks rolled in.

RFE was without question a major factor in bringing the Hungarian people to an uprising, and there is ample evidence that RFE broadcasts could have been interpreted as implying that American help would be forthcoming. Subsequently there was a US government investigation which, as might be expected, found mistakes on the part of RFE but no major measure of guilt. The former Director of Engineering Services at RFE in the early 1960s, Russell Geiger, explained the event to me: 'Though RFE was without question a major factor in bringing the Hungarian people along the revolutionary path, it was not the only factor, nor was it likely the match that lit the fuse.'

For more than 20 years both RFE and RL existed free from public scrutiny, operating as covert or surrogate propaganda broadcasters behind the heat shield of the CIA. With the passing of the US Freedom of Information Act in 1971, which had the effect of bringing the activities of RFE and RL more into the public domain, the CIA lost control. Both broadcasters were merged into a single entity and placed under the administration of the Board for International Broadcasting.

However, the political objectives reminded the same, and funding came from the American taxpayer through unpublicised channels. Propaganda broadcasts became less bellicose and more refined in character. The unqualified success in helping to bring about the tearing down of the Berlin Wall, in which RIAS played a key role in tandem with RFE/RL, was the prelude to the eventual overthrow of communist ideology and the collapse of the Soviet Empire, one year later.

The RFE/RL transmitter network

It was the practice of RFE/RL engineers constantly to update, hone and refine the targeting ability of its transmitting stations. By the time the cold war had ended its six high power SW sites had a total of 57 SW transmitters with a combined power of 12,400 kW. In comparison with either VOA or the BBC-WS, on a regional basis, RFE/RL was more powerful and arguably more effective than either. This was because the total kilowatt power of RFE/RL was targeted to just two regions; Eastern Europe and the Soviet republics, whereas the resources of VOA and the BBC were of necessity thinly spread to cover the whole world.

The merger of RFE and RL brought advantages as it became possible to have a transmitter site service a number of countries or regions. For example, each transmitter site had the flexibility to target different countries and regions, so that by 1991 the USSR was being covered 24 hours a day by as many as 50 RFE/RL SW transmitters.

The main target countries in Eastern Europe were Bulgaria, Czechoslovakia, Hungary, Poland and Serbia and Croatia in the former Yugoslavia. Targets in the USSR ranged from the Baltic States of Estonia, Latvia and Lithuania, to the Soviet Republics in Western USSR and on to the Republics of the Eastern USSR.

To make it possible to target these countries and regions each of the six SW transmitter sites had directive antennas beamed onto a number of azimuthal bearings, with the ability to slew in azimuth by as much as ±25 degrees. Additionally, most of the curtain arrays possessed the ability to slew the vertical angle of wave departure. For instance, some of the curtains of the 6 vertical stack type used by RFE/RL could give departure angles down to a just-above-ground grazing (2–3 degrees).

Such features gave RFE/RL engineers the ability to optimise propagation paths for different times of day and for differing path lengths and these measures, coupled with innovative antenna designs with the ability to combine as many as four transmitters into a single directive curtain, ensured that SW transmissions from RFE/RL seldom experienced the kind of audibility problems that plagued other broadcasters. The need for high signal audibility in the target zones was driven by the Soviet jammers.

To facilitate easy recognition of a Radio Free Europe or Radio Liberty broadcast, each of the twenty or so language broadcasts opened and closed the programme with a distinctive folk song. Polish broadcasts, for instance, opened and closed with the Polish patriotic song 'Hail Glorious Dawn of May', whilst Russian programmes used another highly appropriate song, 'Hymn to Freedom'.

Table 15.2 *RFE/RL transmitter network*

Site	Location	SW transmitters	Antennas
Biblis	Germany	10 × 100 kW	10 curtain, 2 rhombic
Lampertheim	Germany	9 × 100 kW	6 curtain, 1 dipole
Holzkirchen	Germany	4 × 250 kW	4 curtain
		[1 × 150 kW MW]	
Gloria	Portugal	19 × 250 kW, 2 × 50 kW	21 curtain, 10 rhombic
Maxoqueira	Portugal	6 × 500 kW	6 curtain
Pals	Spain	6 × 250 kW	9 curtain

A more detailed description of RFE/RL's transmitter network can be found in Appendix 1; it is summarised here in Table 15.2.

The total RFE/RL transmitter network comprises some 57 SW and one MW transmitters. The total kilowatt power is 12,400 kW, with a total antenna compliment of 56 high gain curtain arrays, 35 high gain rhombic antennas, all for target service in SW. MW antennas include one dipole with reflecting screen and a four-tower directional antenna.

The dependence on transmitters with 100 kW power at Biblis and Lampertheim is due to the fact that these stations were conceived when 100 kW was the most popular size, before the advent of 500 kW transmitters such as those installed at the newest SW station at Maxoqueira, but also because the distances to target areas from these sites in Germany was not great.

RFE/RL engineers succeeded in achieving phenomenally high signal strengths in target areas due to two important factors. First, the knowledge and understanding which RFE/RL engineering staff acquired and brought to bear on the science of the best angles of wave departure and optimum choice of operating frequency for time of day. Second, the development of high gain curtain arrays and extremely powerful rhombic antennas by the station. RFE/RL was the first broadcaster to bring innovative curtain antennas into service with features such as six-stack high curtains with slewing facilities both in azimuth and elevation, enabling accurate targeting of broadcasts, and diplexing and triplexing for combining transmitters into one directive antenna.

Chapter 16
The restructuring of US Government international broadcasting

One of the important roles of the President's Task Force on US Government international broadcasting, formed on 29th April 1991, was to advise on the most appropriate structure under which all US Government (USG) international broadcasting assets and activities should be consolidated. At that time USG international broadcasting comprised the following entities organised under the administration of the Bureau for International Broadcasting, itself under the USIA.

- Voice of America
- Worldnet TV
- Office of Cuban Broadcasting
- RIAS (the US propaganda radio station in the former western sector of Berlin)

In parallel with this structure, a separate broadcasting body, the Board for International Broadcasting, was the Washington office for Radio Free Europe/Radio Liberty (RFE/RL). From the outset, given the contrasting nature of the broadcasters under the USIA, and the surrogate broadcasters under the BIB, it was evident that the Task Force had been given a difficult challenge. VOA is the official broadcaster representing America to the world. RFE/RL was different in character and in style and its role was to appear to be a European broadcaster speaking for the peoples in the former Soviet Union and the Eastern bloc countries. VOA is funded directly by the US government, fully accountable to Congress and open to public scrutiny.

RFE/RL on the other hand was a surrogate broadcaster, not accountable and not subject to Congress or public scrutiny, run in the same way as a private broadcaster and free to devise, plan and implement projects and missions.

The Task Force duly evaluated all possible options for re-structuring and gave an opportunity to the USIA, VOA and RFE/RL each to make their case. In the summary to the President the Task Force recommendations were to the effect that VOA should stay as part of the USIA, that the operations of RFE/RL should stay under the Board for International Broadcasting (BIB), and that the Office of Cuban Broadcasting (Radio and TV Marti) should also be put under the BIB.

The US Government made its final decision under a new International Broadcasting Act which became law in April 1994. This Act called for a radially new structure, with a single transmission authority to plan and devise an integrated, combined international broadcasting relay network from the two separate entities of VOA and RFE/RL.

In January 1997 the implementation plans were made known, in effect creating the International Broadcasting Bureau, administered by the USIA. The Bureau for International Broadcasting and the Board for International Broadcasting, which administered VOA and RFE/RL respectively, ceased to exist. There was now one transmission network used by both VOA and RFE/RL (see Figure 16.1 and Table 16.1).

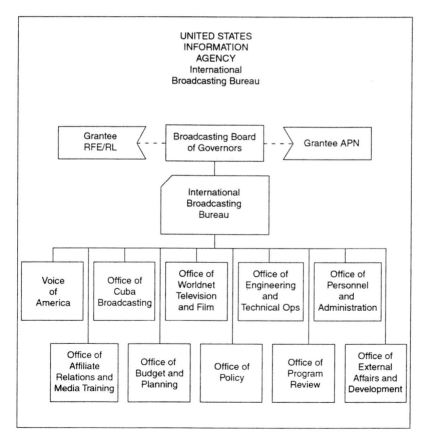

Figure 16.1 *Structure of US Government international broadcasting operations*

Table 16.1 *International Broadcasting Bureau operational and planned broadcast transmitter network (February 1997)*

Current relay stations	Manufacturer/ model	Broadcast transmitters		Total power	Age (years)	Comments
Belize	Harris VP-100B	2 – 100 kW	MW	201.5 kW	9	
	Broadcast Electronics BE-1.5B	1 – 1.5 kW	FM		6	
Botswana (Moepeng Hill)	Marconi B6030	1 – 500 kW	MW	900 kW	2	
	Continental 418E	4 – 100 kW	SW		5	
Colombo (Sri Lanka)	Collins 207B-1	2 – 35 kW	SW	115 kW	46	4 – 500 kW SW and 2 – 250 kW SW to be installed
	Philips	1 – 10 kW	SW		44	
	Collins 207B-1	1 – 35 KW	SW		46	Used by host govt.
Delano (California)	Collins 821A-1	3 – 250 kW	SW	1750 kW	34	
	BBC SK53 C3	4 – 250 kW	SW		16	
Germany Holzkirchen	Continental 318.5D	1 – 150 kW	MW	1150 kW	12	
	Continental 419E	4 – 250 kW	SW		9	
Ismaning	Thomcast TMW 2300 Series 7	1 – 300 kW	MW	300 kW	2	
Lampertheim	Continental 418D	2 – 100 kW	SW	900 kW	20	
	Continental 418D-1	2 – 100 kW	SW		17	
	Continental 418D-2	3 – 100 kW	SW		6	
	Thompson-CSF TRE2315	2 – 100 kW	SW		4	
Greenville (North Carolina)	Continental 420A	6 – 500 kW	SW	6500 kW	44	Schedule reduced at Site B because of budget constraints
	Continental 420B	1 – 500 kW	SW		11	
	GE 4BT250A1	6 – 250 kW	SW		36	
	Marconi B6127	1 – 500 kW	SW		11	
	BBC SK55 C3-2P	1 – 500 kW	SW		11	
	AEG S4005	1 – 500 kW	SW		11	

Location	Equipment	Units	Band	Power	No.	Notes
Kavala (Greece)	Continental 105B	1 – 500 kW	MW	3600 kW	44	Back-up
	Harris DX-600	1 – 600 kW	MW		1	
	Continental 419D	9 – 250 kW	SW		27	Used by host govt.
	Continental 419D	1 – 250 kW	SW		27	
Kuwait	Continental 317C-2	1 – 50 kW	MW	650 kW	16	Back-up
	Marconi B6043	1 – 600 kW	MW		3	
Morocco	Marconi B6-128	10 – 500 kW	SW	5000 kW	3	
Pals (Spain)	Continental 420A	4 – 250 kW	SW	1500 kW	35	
	Marconi B6131	1 – 250 kW	SW		8	
	GE 4BT250B1	1 – 250 kW	SW		32	
Philippines Poro	Harris DX-1000	1 – 1000 kW	MW	1420 kW	1	
	GE 100C	2 – 100 kW	SW		43	
	Collins 207B	2 – 35 kW	SW		46	
	Gates HF-50C	3 – 50 kW	SW		36	
Tinang	Hughes HC-114	10 – 250 kW	SW	3150 kW	29	
	Gates HF-50C	3 – 50 kW	SW		36	
	BBC SK53C3	2 – 250 kW	SW		16	
Rhodes (Greece)	Continental 105B	1 – 500 kW	MW	600 kW	44	1 – 600 kW MW to be installed
	Gates HF-50	2 – 50 kW	SW		36	
São Tomé	Harris MPS10645-00001	1 – 100 kW	SW	1120 kW	1	Tropical Band
	Thomcast AG TMW 600	1 – 600 kW	MW		1	
	Thomcast AG TSW 2100	4 – 100 kW	SW		1	
	Harris MPS 106953-00001	1 – 20 kW	SW		1	Used by host govt.
Thailand, Bangkok	Harris DX-1000	1 – 1000 kW	MW	1000 kW	1	
Udorn	Marconi B6-128	7 – 500 kW	SW	3500 kW	2	One used by host govt.
Total		1 FM, 13 MW, 107 SW		33356.5 kW	Avg. 18	
Planned relay station						
Tinian		3 – 500 kW	SW	1500 kW		

Analysis of this data shows some interesting facts. Some of the RFE/RL SW stations will cease to exist and in fact the demolition has already begun. The dismantling of these SW stations is all the more perplexing because of their uniqueness. Gloria by itself was the largest SW station in the western hemisphere with 21 high power SW transmitters.

Equipment from the decommissioned stations, transmitters, feeders, antennas and switching matrices, have been shipped to other SW stations, one of which is Tinian Island in the Northern Marianas, a fact which substantiates that the closure of the RFE/RL sites was not because the equipment was outdated. Had the USIA so wanted, there were several good reasons for keeping these former sites intact, such as leasing them out, as Russia does with its jammer sites. Another would be to turn over the sites as a good will gesture to Portugal as pay-back for being host country, a practice to which the US State Department subscribes (as is evident from the fact that some former VOA transmitters hosted in Greece are to be given to the Greek Government).

Yet another solution for the sites at Gloria and Maxoqueira would have been to mothball them for future contingencies. It is rare for SW sites to be abandoned and the US President's Task Force noted in its report 'We are living in an uncertain world and America needs the capacity to reach out to potential hot spots around the world'. The closure of these RFE/RL sites may have more to do with other things. Sometimes it is politically expedient for governments to play down and even conceal true facts of war, and the cold war was a war in which RFE/RL played a controversial role.

There has already been extensive press coverage in the USA of the achievements of VOA, particularly during and after the abortive coup of 19–21 August 1991 during the Moscow uprising. This was when Boris Yeltsin's staff sent a message to Washington by fax that his appeal to the army should be transmitted over VOA transmitters to Russia. 'Do it now' was the request. The appeal generated good publicity for VOA and helped to create an image that it was the broadcasts from VOA that had been the primary tool for the projection of news which brought succour and hope to the supposedly oppressed peoples of the Soviet Union.

In fact VOA broadcasts to Russia seldom achieved the audibility level of the BBC World Service, or those from RFE/RL whose penetration rate of Russian jamming was on a scale that surpassed any other broadcaster. According to one of its regular listeners in Moscow with whom I spoke, there never was an audibility problem with Radio Liberty; 'It came in loud and clear, just like a local radio station'. Quite evidently if Yeltsin's aides had been better informed they would have sent the request to Radio Liberty!

The SW transmitter infrastructure of RFE/RL at six sites in Europe constituted a network unequalled by any other broadcaster. For this reason I believe some of these SW transmitter sites should have been permitted to exist for the benefit of future generations, in the same way that some wartime communication bunkers have been turned into museums. They could also be a source of information on the tools used in the cold war, which would benefit historians.

Table 16.2 of stations represents the position for IBB sites and transmitters in April 1998 and updates Table 16.1.

Table 16.2 *Status of Major Construction Projects IBB transmitter sites, April 1998*

Bangkok Transmitting Site *(Thailand Relay Station):* A solid state 1000 kW MW transmitter became operational in May 1996.

Belize Relay Station: This facility with two 100 kW MW transmitters became fully operational in July 1990. A 1.5 kW FM stereo transmitter was added in February 1995.

Bethany Relay Station *(Ohio):* This SW station was closed in November 1994 and has been dismantled.

Botswana Relay Station: In December 1991, the station's four 100 kW SW transmitters began broadcasting from the Moepeng Hill site. In June 1994, Botswana's MW facility with its 600 kW transmitter went on the air to upgrade an existing 50 kW transmitter. A solid state modulator will be installed in the MW transmitter this summer.

Delano Relay Station *(California):* Solid state modulators have been installed in four 250 kW SW transmitters.

Germany Relay Station *(Biblis Transmitting Site):* Solid state modulators are being installed in two 100 kW SW transmitters of this former RFE/RL facility; the station is now remotely operated from the IBB's Lampertheim transmitting site.

Germany Relay Station *(Holzkirchen Transmitting Site):* This former RFE/RL station has had solid state modulators installed in two of its four 250 kW SW transmitters; the remaining two solid state modulators will be installed by the end of this calendar year. The station is now remotely operated from the IBB's Lampertheim transmitting site.

Germany Relay Station *(Ismaning Transmitting Site):* A solid state 300 kW MW transmitter became operational in March 1995.

Germany Relay Station *(Lampertheim Transmitting Site):* Four new antennas have been installed and are undergoing testing at this former RFE/RL SW transmitting site. Five of the site's nine 100 kW SW transmitters have been equipped with new solid state modulators. A new station automation system also has been installed.

Gloria Transmitting Site *(Portugal):* This former RFE/RL facility ceased broadcasting in May 1996. The station's transmitters and switch matrix are being disassembled and readied for shipment to IBB sites in Greece, Playa de Pals (Spain), and possibly other IBB network locations.

Greenville Relay Station *(North Carolina):* In March 1995, the receiver site and office area were closed and consolidated into one of the station's two transmitting sites.

Kavala Transmitting Site *(Greece Relay Station):* A solid state 600 kW MW transmitter became operational in December 1996. Two 250 kW SW transmitters from the closed Gloria facility are being installed in Kavala with new solid state modulators. Two new low band antennas also are being added to the transmitting site. In addition, the switch matrix has been expanded and has received a new automation system.

Kuwait Relay Station: In May 1996, the permanent facility began broadcasting from a new 600 kW MW transmitter.

Maxoqueira Transmitting Site *(Portugal):* This former RFE/RL facility ceased broadcasting in September 1994; almost all of the transmission equipment other than antennas has been shipped to Tinian for installation.

Morocco Relay Station: This state-of-the-art facility with its 10 500 kW SW transmitters went on the air in October 1993.

Pals Transmitting Site *(Spain Relay Station):* Current plans call for this former RFE/RL station to receive more modern transmitters and the switch matrix equipment from the closed Gloria facility over the next several years.

Poro Transmitting Site *(Philippines Relay Station):* In June 1996, a more efficient 1000 kW MW transmitter went on the air to Southeast Asia. The IBB is investigating alternate sites for this transmitter.

Rhodes Transmitting Site *(Greece Relay Station):* The IBB is installing a solid state 600 kW MW transmitter in a new facility. The older, dual SW/MW facility will be closed; SW broadcasting from Rhodes will be terminated; and a new MW facility will be remotely controlled from Kavala. In addition, the site's SW antennas will be relocated to Kavala.

São Tomé Relay Station: Completion of this permanent relay station with its five 100 SW transmitters and one 600 kW MW transmitter in April 1996 improved IBB broadcast coverage to almost two-thirds of the African continent.

Sri Lanka Relay Station *(Under construction):* Although the station was scheduled to go on the air in 1995, a series of calamities ranging from a fire in 1996, to design, material, and installation problems with the towers over the past several years, to continual civil strife have delayed completion of the new facility. The station may become operational by the end of calendar year 1998. Consideration is being given to increasing the capability of this strategically important site.

Tinang Transmitting Site *(Philippines Relay Station):* New solid state modulators have been added to three of the site's 15 250 kW SW transmitters. Three of the site's oldest SW transmitters are being replaced with newer transmitters shipped from the closed Bethany Relay Station. These newer transmitters will be equipped with solid state modulators.

Tinian Relay Station *(Under construction):* This new station is located on the Island of Tinian in the US Commonwealth of the Northern Marianas. A new building, power plant, and utility system have been built. Transmitters, switch matrix, transmission lines, and control room equipment from the closed Maxoqueira facility are being installed. New antennas capable of surviving 250 kilometer per hour wind speeds are being erected. Threre transmitters are expected to being broadcasting in early 1999, with three more transmitters expected to come on line about one year after that.

Udorn Transmitting Site *(Thailand Relay Station):* With its seven 500 kW SW transmitters and 25 curtain antennas, the station can reach over 40 per cent of the world's population. Completion of this facility in October 1993 achieved a major goal of the IBB's modernisation program.

Chapter 17
The Arab–Islamic world

In the seventh century AD the Arabs, united by a single tongue and a single faith, Islam, created a vast empire which at its peak extended from the Pyrenees to the Magriab of North Africa, along to Egypt, the Arabian Gulf and deep into central Asia. This mighty empire lost its unity and its independence over time as one Arab state after another lost its independence to the British, the French and the Italians. Some became colonies of the French and the Italians whilst others became protectorates under the British. Today the Arab-Islamic world has recovered its independence but it can never be the great empire it once was.

Nevertheless the Arab-Islamic world of today is unique. It is a collection of states which are bonded by a common heritage, a common culture, a common faith and a common tongue. It is because of these qualities that the Arab-Islamic nations have invested so much in the construction of super-power MW transmitters for the purpose of long-range regional broad-casting. The region extending from Morocco to the Gulf and beyond, has more high power and superpower MW transmitters than any other part of the world.

Nevertheless this high investment in MW and LW transmitters has not been at the expense of global broadcasting in the high frequency spectrum. It is significant that over the past two decades investment by a number of Arab-Islamic nations in the development of global broadcasting capability has exceeded that of many Western powers. Sales of 500 kW SW transmitters to the Arab-Islamic world supply ample proof of this phenomenon. Table 17.1 summarises these acquisitions for nine countries. When account is taken of the population figures for some of these the sheer scale of investment into high power SW becomes apparent.

This table excludes the SW transmitter sites of Voice of America and BBC World Service which exist in certain countries. For example, Morocco is host to the VOA SW site which has ten 500 kW SW transmitters. Iran, although not an Arab state, is an Islamic state and is included here due to its political and economic importance to this region.

137

Table 17.1 *Arab-Islamic and Iranian investment in high power (500 kW) SW transmitters*

Country	Population (millions)	Region	No. of 500 kW transmitters
Iran	65.0	Middle East	26
Iraq	21.9	Middle East	22
Kuwait	2.0	Gulf	21
Libya	4.7	N Africa	11
Turkey	62.5	Near East	9
Saudi Arabia	17.5	Arabia	6
Egypt	58.2	N Africa	5
Arab Emirates	2.2	Gulf	4
Qatar	0.6	Gulf	1
Totals	234.6		116

Note: Figures for 500 kW transmitters are correct until 1998 although not all are necessarily in service.

Broadcasting with superpower from the Arab world

The Arab culture is unique in many ways, one of which is that it is primarily an oral based culture. Up to the present day the spoken word takes precedence over the written and it is only in the past thirty years that the Islamic countries have moved towards a written contract of agreement. Given the importance that is attached to the spoken word it is easy to understand why radio broadcasting has become so popular, since it was first introduced in the 1940s. From an entertainment aspect, no tea house or café is complete without a radio set rendering the captivating Arabic music which might be coming from a powerful MW broadcasting station several hundred miles away. Moreover it has at all times been a powerful tool enabling Arab leaders to reach out to their peoples.

A key reason for the popularity of high power AM broadcasting is that there is a much higher degree of homogeny to be found in the Arab world. United by a common tongue, it is quite common to find Egyptian nationals working in other Arabic countries and the same applies to many other Arab nationalities. These expatriots love to hear the voice of their homeland over the airwaves. High power and superpower AM does just that.

A 600 kW MW transmitter can project a signal over a long distance during the day and even further at night, taking advantage of ionospheric reflection. A 2000 kW carrier power output into a six-in-line directive antenna can bridge vast deserts and empty quarters, crossing international frontiers. The common language and culture simplify programming and mean that programmes can be enjoyed with equal enthusiasm by listeners in any Arabic-speaking country.

Most Westerners who have lived or worked in the Arab world quickly become familiar with listening to Arabic radio stations on the medium waveband, especially after dark when it is alive with programmes that could be sent from the other side of the Arab world. But perhaps the best proof of the popularity of AM broadcasting is to be found in an analysis of broadcasting power. From Morocco eastwards to Iraq there are no less than nineteen high power and superpower AM stations with carrier output powers of 1500–2000 kW, despite the relatively low populations. The Arab state of Qatar, for example, is the smallest independent Arab state in area, with a total population equal to a suburb of New York, Paris or London, yet its Ministry of Information and Culture has a broadcast power far exceeding that of any other nation when measured on a kilowatt per capita basis. Kuwait and the United Arab Emirates are in the same league, each with a population of around two million citizens.

The Islamic countries in North Africa and the Middle East total more than twenty nations. Table 17.2 summarises the capacity in AM broadcasting of the 12 most significant of these. Table 17.3 shows superpower transmitters, where seven of the countries possess at least one 2000 kW AM transmitter, It should be noted that the transmitter capacities in both tables exclude those transmitting stations which are part of the worldwide network of VOA, RFI, the BBC-WS or any other foreign broadcaster.

The most powerful Middle East broadcaster on the medium waveband is Saudi Arabia, followed by Iraq and Iran, respectively. However, if account is taken of the fact that Iraq had much of its broadcasting capacity damaged in the Gulf War and later attacks, whilst Iran was actually increasing its broadcasting capacity, then the placings become reversed.

Table 17.2 *Estimated AM broadcasting capacity of the more significant states in the Arab region*

Country	Receiver count	MW/LW (kW)	SW (kW)	Total kW capacity
Iran	13 000 000	12 540	10 000	22 540
Iraq	3 700 000	11 000	10 000	21 000
Saudi Arabia	3 800 000	15 000	5500	20 500
Libya	1 000 000	5800	3000	8800
UAE	490 000	5300	4200	9500
Kuwait	1 000 000	3700	7500	11 200
Jordan	980 000	4400	2000	6400
Morocco	5 000 000	6200	850	7050
Egypt	16 450 000	5000	4800	9800
Syria	3 000 000	3500	1500	5000
Algeria	3 500 000	7800	500	8300
Qatar	180 000	4000	0	4000

Note: Figures for broadcasting capacity are for 1992, since then some expansion has taken place. The source of figures for receiver counts is the WRTH Handbook 1995.

Table 17.3 *Superpower AM stations in North African and Middle East Islamic states, 1990*

Country	Location	Carrier power	LW/ MW	Supplier
Algeria	Ouargla	2000 kW	LW	Asea Brown Boveri
	Bechar	2000 kW	LW	Asea Brown Boveri
	Tipaza	1500 kW	LW	Asea Brown Boveri
Morocco	Nador	2000 kW	LW	Thomson-CSF
Iraq	Tanaf	2000 kW	MW	Thomson-CSF
	Missan	2000 kW	MW	Thomson-CSF
	Sulaimaniya	2000 kW	MW	Thomson-CSF
Saudi Arabia	Dubai	2000 kW	MW	Continental
	Qurayyat	2000 kW	MW	Continental
	Jeddah	2000 kW	MW	Continental
	Dammam	2000 kW	MW	Continental
	Jeddah	2000 kW	MW	Continental
Jordan	Ajlun	2000 kW	MW	Continental
Egypt	Batrah	2000 kW	MW	Continental
UAE	Abu Dhabi, Dabiya	2000 kW	MW	Asea Brown Boveri
	Abu Dhabi, Dabiya	2000 kW	MW	Asea Brown Boveri
	Abu Dhabi, Sadiyat	1500 kW	MW	Asea Brown Boveri
	Dubai	1500 kW	MW	Marconi
Qatar	Doha, Al-Arish	1500 kW	Mw	Marconi

Note: This table does not include 2000 kW installations in Libya which are not in operation. Figures are based upon data compiled by author. The capability of Iraqi installations is unclear due to ongoing military action.

Iran is continuing to expand its MW broadcasting capacity and if the pace continues it could become the most powerful broadcaster in the Middle East within the next few years.

All the signs show that AM broadcasting in this region of the world is, notwithstanding the emergence of new technologies, assured of a healthy growth rate for several decades to come. No other form of communication can with the same ease fulfil the four vital roles that high power AM does; bringing kings and leaders in touch with the people, a tool during national disasters, entertainment, and a link with workers in adjacent Arab states.

SW broadcasting

Arab interest in the use of the broadcast bands in the HF spectrum began in the 1960s. From a slow start the pace accelerated with increasing oil revenues, and by the late 1980s vast revenues were being assigned to the acquisition of powerful SW transmitter sites. Indeed, for some of these Middle East countries their enthusiasm for superpower broadcasting in the medium waveband was even exceeded by a desire to possess the most powerful and the most sophisticated 500 kW SW transmitters. Three of the Middle East states stand out in particular. These are Iran, Iraq and Kuwait, whilst a fourth country, Jordan, is worthy of mention because of the uniqueness of its installations.

Kuwait was the first to invest in the latest 500 kW transmitters; it purchased ten over a period of a few years to 1989. Also in the mid 1980s its neighbour Iraq awarded one of the biggest SW contracts ever to Thomson-CSF. Amongst other equipments, the contract included a SW station at Balad to be equipped with sixteen 500 kW transmitters and ninety-eight SW antennas. When this station was completed it was the most powerful SW station in the world.

When Iraqi forces invaded Kuwait in 1990 they took control of Kuwait's SW transmitting station and apparently liked the 500 kW transmitters which Kuwait had bought from Asea Brown Boveri so much that they stole a pair, so the story goes, and took them back to Baghdad. The situation presently with Iraq's SW transmission capability is not clear; some were damaged in the Gulf War, and likely in more recent US/UK attacks, and others held up for spares. Clearly this period has put paid to Iraq's ambition of being the most powerful Islamic SW broadcaster.

That mantle has now been taken over by its neighbour Iran. From the early 1980s Iran has had a powerful capability on the short waves. Four high power SW sites at Kamalabad, south of Tehran, Ahwaz in the south and two more in the eastern sector, Mashhad and Zahedan, are equipped with 250, 350 and 500 kW transmitters mainly supplied by the Swiss company ABB. The fifth and most recent high power SW station in Iran is at Sirjan where construction started in 1990. Constructed by Telefunken Sendertechnik, it is equipped with ten 500 kW SW S4105 transmitters, the newest SW transmitter from Telefunken. These additional ten transmitters bring Iran's SW broadcasting capacity to the top of the Middle East league, and also rank it as one of the most powerful in the world.

It is true to say that apart from the high sales of SW transmitters to VOA, it is the Middle East states which have sustained the high power transmitter market since the late 1980s. Their support for the three[1] big European transmitter manufacturers ABB, Telefunken and Thomson has enabled these companies to go on to develop even better and more advanced 500 kW SW transmitters.

---

[1] In 1993 Asea Brown Boveri sold is transmitter manufacturing business ABB Infocom to Thomson-CSF, to become part of a new Thomson company called Thomcast (see Chapter 23).

Figure 17.1 *Balad SW station, Iraq, with 16 Thomson-CSF 500 kW transmitters*

An assessment of the MW transmitter markets in Arab countries

In securing the contract to re-equip the tiny but wealthy oil state of Qatar with a 2000 kW MW transmitter system, the US company Harris has tapped a potentially large market: the replacement of now ageing, high power AM transmitters, some of which are of up to 2000 kW carrier power. There are two main regions of the world where such markets exist. One is the CIS where the former republics of the USSR possess several hundred, but their severe economic difficulties will probably kill prospects for several more years.

The other region is the Arab-Islamic world, where many countries began investing in radio broadcasting from the 1950s, with the pace

Figure 17.2 *500 kW SW rotatable curtain array antenna, Balad, Iraq*

quickening by the 1970s. Investment has been in two phases, initially with medium power and then moving to the acquisition of 1000 and 2000 kW transmitters. Now many of these are approaching a stage when it will be economically beneficial to replace them with modern energy-saving transmitters, whilst at the same time obtaining better performance. The cost of running a two megawatt transmitter is very high – it equates to the electrical power consumption of a small town – so power-saving is obviously important.

Traditionally MW and LW transmitters tend to say in operational service for much longer than SW transmitters, one of the chief reasons for which is that there have been fewer technological developments in AM transmitters. However, the introduction of modern, high efficiency AM transmitters in the 1990s, with much increased overall electrical conversion (RF power output/AC input power) has now made it desirable to replace old and inefficient high power AM transmitters.

Replacing an old class B type transmitter of 1960s vintage, which has an efficiency of about 60–63 per cent with a modern energy-saving transmitter would yield a saving of about 22 per cent on electrical consumption. With a saving of this order it has been calculated that the pay-back on energy costs would be sufficient to cover the cost of a new transmitter over a period as low as five years.

The potential market for new AM transmitters in the North Africa–Middle East region alone has been estimated at about 100 for those with output powers from 500 kW and upwards, and the manufacturers who will benefit from this market include Continental Electrics, RIZ, Telefunken, Nautel, Harris, and Thomcast.

Broadcasting with SW superpower from the Arab-Islamic world

This section is a distillation of some earlier extensive studies into radio broadcasting from certain North African and Middle Eastern countries, which began when it became apparent that many of these countries had begun to make significant investments in radio broadcasting using super-power. Radio broadcasting came new to the Arab world of North Africa and the Middle East from the early 1950s as a result of these countries coming into close contact with Allied and German armies during World War II. There were problems, however; radio transmitters had to be bought from the West but many Arab countries at that time did not have the financial resources to pay for such imports. In the end the funding came

Figure 17.3 *SW/MW/LW transmitting station at Qasr Kherane, Jordan*

from the USA, the Soviet Union and Britain in exchange for things like military bases.

Newer and better educated Arab leaders were quick to realise that radio broadcasting was much more than a social tool for entertainment. It had potential as a powerful tool for keeping in touch with the masses and as such they saw it as a tool of government and one that should remain firmly under government control at all times. To this end Arab states passed decrees to ensure that radio broadcasting should be under state control.

The transformation of many Middle East states from being poor to very rich began in the early 1970s. Thus, apart from a few countries such as Egypt and Jordan, many states began to build huge wealth from oil resources. By the early 1970s the acquisition rate of high power and super-power broadcasting facilities began to soar when these very wealthy Arab states, who belonged to the OPEC club, realised that they needed a strong voice in international affairs.

It is significant that the first countries to invest in the latest generation of high power SW transmitters – the so-called 500 kW super transmitters – were these Arab-Islamic states. A survey by the author in 1990 showed that the combined total broadcast power of a dozen Arab states was as great or greater than that of a major Western country, and with superior technology.

Whatever the outcome of the long-running, on-off war between the US and Iraq, one thing remains certain – the Arabic world will never be what it was. These countries will want to exercise control over their own destiny and what better way of achieving that end than to use the very same tool that Western governments used to help bring about the end of Soviet communism?

Market leaders in the supply of broadcast infrastructure

The Aramic–Islamic world was the region most assiduously courted by the Swiss company Brown Boveri (BBC). The French, with their former colonial interests, also achieved success, but never on the same scale as the Swiss manufacturer. Some analysts might ascribe Swiss success to Swiss neutrality but this would be a too simplistic analysis. Good political relations can help secure an initial contract but do not account for repeat orders. In the final analysis these Arabic–Islamic countries place greater emphasis upon having the latest technology and a guarantee of reliable operation, coupled with dedication to after sales service. The Arab-Islamic world is a region where rewards can be high but at the same time where a reputation can easily be lost if a company fails to meet its contractual obligations.

Table 17.4 *Analysis of sales of 100, 250 and 500 kW transmitters 1970–1995*

World region	Transmitter Power kW	Percentage market share					Total no. sold
		ABB	Thomson	Telefunken	Marconi	Continental	
Arab	500	53.8	34.6	9.6	0.9	0.9	104
Arab	250	51.8	29.6	0	16.6	1.9	54
Arab	250+500	54.2	33.5	6.5	4.5	1.3	158
Rest of World	500	24.0	22.6	22.6	19.3	11.3	150
World	500	36.2	27.5	17.3	11.8	7.0	254
World	250	35.8	14.8	7.0	27.9	14.4	229
World	250+500	35.9	21.4	12.6	19.8	10.5	484
World	100	25.8	16.5	12.7	2.1	42.7	236
World	100–500	34.0	19.7	12.7	13.8	19.7	719

Analysis of sales for SW high power 1970–1995

Tables 17.4 and 17.5 tabulate the total sales of different power transmitters (100, 250 and 500 kW) to the Arab-Islamic region, compared with total world sales. The market percentage for each of the five major companies is expressed as a percentage of the total for each category of transmitter power. The tables analyse the period 1970–1995 and with it 1985–1995, respectively.

Table 17.5 *Analysis of sales of 100, 250 and 500 kW transmitters 1985–1995*

World region	Transmitter power kW	Percentage market share					Total no. sold
		ABB	Thomson	Telefunken	Marconi	Continental	
Arab	500	48.7	35.8	12.8	1.3	1.3	78
Arab	250	83.8	16.6	0	0	0	6
Arab	250+500	50.6	35.8	12.3	1.3	1.4	84
Rest of World	500	30.0	14.4	10.0	26.6	19.6	90
World	500	37.9	24.6	11.4	15.0	10.8	168
World	250	43.2	0.9	0.3	33.0	11.4	88
World	250+500	39.8	20.0	8.6	21.3	11.0	256
World	100	33.3	11.4	0	0	55.2	96
World	100–500	39.6	16.8	6.0	15.0	22.4	362

Some of the conclusions that can be drawn are as follows:

- The Arab-Islamic world accounts for 41 per cent of world sales at 500 kW (49 per cent during 1985–1995) and 23.6 per cent at 250 kW (6.8 per cent in 1985–1995)
- The most popular size of transmitter was 500 kW
- The market leader in the Arab-Islamic region ws ABB, followed by Thomson (ABB was world market leader, except at 100 kW where Continental led with 42.7 per cent)

Eleven of the 22 Arab-Islamic states possess a SW broadcast capacity, measured on a kilowatt per capita basis, greater than that in Western countries.

In conclusion, this region is a growth market for SW, especially in the 500 kW category. It is a market which, by virtue of its size and demand for the latest technology, commands special attention for the broadcast equipment manufacturers.

Chapter 18
Libya, Egypt, Kuwait and Iran

Libya

Libya was once a fashionable resort for the kings and queens of Europe. When it was taken from Italy after World War II and placed under British administration, oil flowed to swell the coffers of the US and Britain, but not the stomachs of the Libyan people. On 1st September 1969 Muammar al Qaddafi changed all that. The puppet King Idris was overthrown and Libya became the newest Arab state. Qaddafi and his revolutionary colleagues represented something entirely new in Arab rulers: a new educated elite, well read in many Western languages. The young, vigorous 28 year old colonel was fired with ambition to make Libya a truly twentieth century state; his first five-year plan achieved all the things he had promised his people. For the first time there was education for all and the quality of life was improved dramatically.

Nevertheless, Western powers (notably the US and Britain) did not share his aims and ambitions and in order to combat Qaddafi, who was seen as a threat to American colonialism and British imperialism, embarked upon a policy of vilification and character assassination. He was portrayed as a dangerous fanatic, and propaganda such as this, with Western media assistance, mostly achieved its objectives. As an engineer who visited Libya before and after the revolution, however, the author witnessed first-hand the transformation in quality of life, the people and the building of schools and universities.

But propaganda can work work both ways, and four years after the revolution Qaddafi embarked upon a project to make Libya a powerful voice on the international airwaves. By 1973 Libya had acquired a number of high power MW transmitters which ranged in output power up to 1000 kW. In 1978 Libya bought a 2000 kW MW transmitter from Radio Industries Zagreb (RIZ) and this was shortly followed by another 2000 kW MW transmitter. It needs to be kept in mind that Libya is a large country in area,

148

with most of the population living in the coastal regions or major cities, so high power MW is the best means of broadcasting national radio to all.

In 1977 Libya began to build up a voice on the short waves as an instrument for projecting foreign policy, but also to enable its citizens in the West and in other parts of the Middle East to maintain contact with their homeland. There are now SW transmitter sites at: Tripoli-Sabrata (13.11 E × 32.54 N), Sebha (14.50 E × 25.52 N) and Benghazi (20.04 E × 32.08 N). Of these sites Sebha and Tripoli are fitted with low to medium power SW transmitters (10–100 kW), mostly installed in the 1970s. The Sabrata site is the most modern and equipped with several 500 kW transmitters (see Table 18.1). However it is likely that the transmitters supplied in 1977 will not now be in regular service. Even so, this leaves Libya, with a population of 4.7 million (less than half that of London) with SW power of seven 500 kW transmitters.

Libyan Jamahiriya Broadcasting External Services calls itself 'The Voice of the Great Homeland'. Programmes open and close with the national anthem and are broadcast in German, Romanian, Hungarian, Polish, Bulgarian, Czech, Slovak and Russian.

Table 18.1 *SW transmitters at the Tripoli-Sabrata site*

SW transmitters and powers	Supplier	Year
4 × 500 kW	Thomson-CSF	1977 (now being decomissioned)
2 × 500 kW	ABB	1983
1 × 500 kW	ABB	1990
2 × 500 kW	Thomson-CSF	1994
2 × 500 kW	Thomcast	1995

Egypt

Egypt was the first Arab country to come under the influence of radio broadcasting, due to the fact that the country had been under the administration of the British long before radio broadcasting came on the scene. Egyptians of today have not lost that love of listening to a radio programme. During the day, but especially after dark, the sounds of Egyptian music pervade the streets of Cairo and Egyptians of all classes are to be seen in restaurants, cafés and tea houses absorbing Arabic programmes.

Not unnaturally Egypt was one of the first Arabic countries to have a powerful radio station. In 1964 a 1000 kW MW transmitter was installed near Alexandria, believed to have been financed under an aid programme by the US. This was followed in 1981 by a 2000 kW MW transmitter. In both cases the contractor was the US company Continental Electronics.

Today both are still operational, though out of deference to age they operate at reduced power.

High power and superpower broadcasting in the MW and LW bands will continue to be used by all the Arabic–Islamic countries as the medium for the projection of foreign policy and other purposes in this region. It was not until the 1970s that the Egyptian government, through its radio and TV authority Egyptian Radio and TV Union (ERTU), moved towards SW broadcasting.

ERTU operates its foreign service from three sites (Table 18.2).

Table 18.2 *ERTU SW stations and transmitters*

Site	Coordinates	SW transmitters and powers	Supplier	Year
Abis	30.05E × 31.10N	2 × 250 kW	Brown Boveri Co.	1976
		2 × 250 kW	Thomson-CSF	1979
		1 × 500 kW	GEC Marconi	1996
Abu Zaabal	31.22E × 30.16N	1 × 100 kW	Thomson-CSF	1990
		2 × 500 kW	ABB	1996
Mokattam	31.15E × 30.03N	1 × 150 kW	CEC	1980

Egypt's foreign service on SW is radiated to all parts of the world. Programme languages include Afar, Albanian, Arabic, Bambara, Bengali, English, French, Fulani, German, Hausa, Hindi, Indonesian, Italian, Lingala, Malay, Oulof, Persian, Portuguese, Pushtu, Russian, Shona, Ndebele, Somali, Swahili, Thai, Turkish, Urdu, Uzbek, Yoruba and Zulu. Dependent on the target region, programmes are preceded by the station identification 'Voice of the Arabs' or 'The Voice of Africa from Cairo'.

Kuwait

If proof is needed of the way the Arab world has taken to international broadcasting in the high frequency spectrum as an instrument for projecting foreign policy, then the Sheikhdom of Kuwait supplies it. Kuwait is a tiny state located at the head of the Arabian Gulf and sandwiched between southern Iraq and northern Saudi Arabia, with a population of 1.59 million. Yet its investment in high power SW transmitters for long range broadcasting, high power MW transmitters for national and over-the-border radio broadcasting and superpower UHF television stations for over the horizon TV broadcasting is on a scale more appropriate to a much larger nation.

Table 18.3 *SW transmitters sited at Kabd, Kuwait*

SW transmitters and powers	Supplier	Year
2 × 250 kW	Brown Boveri Co.	1968
4 × 500 kW	Brown Boveri Co.	1976
2 × 500 kW	Brown Boveri Co.	1979
2 × 500 kW	Brown Boveri Co.	1981
2 × 500 kW	Brown Boveri Co.	1988
6 × 500 kW	ABB	1990/1
2 × 500 kW	ABB	1992
3 × 500 kW	ABB	1995

Kuwait has its main SW site at Kabd (47.55 E × 29.16 N). This station is equipped with SW transmitters as listed in Table 18.3. Those transmitters installed prior to 1988 are not in regular service. Even so, the twelve from 1988 to 1995 are in regular service, a truly impressive amount of SW transmitter power for such a small state.

Kuwait was invaded by the forces of Iraq on 2nd August 1990, and the Emir and his family fled over the border to Saudi Arabia. An Iraqi government administration was formed and Kuwait *de facto* became the nineteenth province of Iraq. Saddam Hussein's claim to Iraq was not without substance because Kuwait had once been part of the Ottoman Pashalik of Basra, but after World War I it became a British protectorate. Saddam Hussein evidently counted on some sort of welcome because in 1986 the Emir disbanded parliament. Kuwait became a dictatorship and the Emir's rule denied citizenship to half of the population because they could not prove fixed residence. This rule affected many who had been born in Kuwait, such as nomadic bedouins. Saddam Hussein was mistaken about the kind of support he could expect, but he was no less mistaken about the reaction of the US and Britain. Appropriation of Kuwait's massive oil wealth sent reverberations through OPEC and the Western world.

In the Gulf War which followed, and the subsequent retreat by Iraqi forces, much damage was done to communications and radio broadcasting installations. The SW complex at Kabd was one of those installations which suffered. From 1991 the balance of the order for six 500 kW SW transmitters awarded by Kuwait in 1990 was completed and a further five 500 kW transmitters were supplied by the Swiss company Asea Brown Boveri from 1992 onwards.

VOA in Kuwait

In the spring of 1996 a new MW transmitter came on air from Kuwait. This was Voice of America's latest expansion in the Middle East. With a carrier power output of 600 kW, the *Wall Street Journal* noted that 'this new transmitter is twelve times more powerful than any domestic radio station in the United States and will prove an invaluable source of uncensored information for the oppressed peoples of Iraq and Iran' (28th May, 1996). The word 'oppressed' always comes into context when Americans talk or write about these two Islamic fundamentalist states. It is a word calculated to arouse emotion amongst Westerners who enjoy pluralistic societies. Yet, in the context of many Iraqis and Iranians it is not as true as when applied to the second class citizens in Kuwait and Saudi Arabia against whom there is discrimination. In fact Iraq is a relatively liberated society (for much of its population, but not all) when compared with Saudi Arabia, as is evident to Western visitors.

VOA's 600 kW MW transmitter reaches out to listeners as far away as India, as well as the Middle East and the Gulf region. VOA likes to believe it has large audiences in these countries. As might be imagined, the Iraqi and Iranian authorities are not pleased about the presence of VOA's broadcasts on MW. The Tehran daily *Jomhuri Islami* called for a war of the airwaves with the State of Kuwait for having allowed the construction and siting of this new transmitter (supplied by GEC-Marconi). Propaganda however can be a double-edged sword, it can provoke a reaction. In this context there are large numbers of Kuwaiti dissidents living in Southern Iran who have fled across the border from Kuwait, so it is possible for Iran to retaliate by broadcasting propaganda to these former citizens of Kuwait.

Though Kuwait is an Arab state, its future, because of its wealth and its past history as as province of Iraq, is tied to American presence and American diplomacy. America needs Kuwait for its oil and Kuwait needs the security of America. So long as this part of the world remains in a permanent state of tension, high power and superpower broadcasting will play a key role in regional and world politics.

The Islamic Republic of Iran

With a population of 65 million and rising (in 1988 it was less than 50 million), Iran can justifiably claim to be the superpower of the Middle East. Iraq, its neighbour has a population of about 22 million. Iran can also claim to be the most powerful broadcaster of all the Arabic–Islamic countries.

Iran has long been a major player on the international AM wavebands. It is both a regional broadcaster by virtue of its powerful MW transmitters and an international broadcaster with a capacity that is second to none in the region when measured by its number of SW transmitters and total kilowatt power. It is important to remember that Iran is not an Arabic nation. The spoken language is Farsi (Persian). It is, nevertheless, an Islamic nation and thus has cultural ties with the Arabic countries.

Iran has a very powerful national radio broadcasting network of more than 70 MW radio stations operating at up to and over 1000 kW, which are scattered throughout Iran and along its borders. Islamic Republic of Iran Broadcasting (IRIB) also has 39 studio centres, outside those in Tehran itself.

In common with a number of Arabic countries, Iran broadcasts foreign language programmes over this national radio network, including Armenian, Bengali, English, French, Pushtu, Urdu, Serbo-Croat, Russian and Spanish.

For its external services IRIB operates some high power MW transmitters but the real strength of its external services is in its SW sites, which tend to be strategically situated on the western and eastern borders of Iran (see Table 18.4).

Table 18.4 *IRIB SW stations and transmitters*

Site	Region	Coordinates	Transmitters	Supplier	Year
Kamalabad	Caspian	51.27E × 35.46N	2 × 100 kW	Telefunken	1963
			2 × 100 kW PSM	ABB	1990
			2 × 250 kW PSM	ABB	1990
			6 × 500 kW Class B	ABB	1984–5
			6 × 500 kW PSM	ABB	1986–8
Mashhad	NE border	59.33E × 36.15N	4 × 500 kW PSM	ABB	1986–8
			2 × 250 kW PSM	ABB	1993
Sirjan	South	56.41E × 29.27N	10 × 500 kW PDM S4005	Telefunken	1990–5
Zahedan	SE border	60.53E × 29.28N	4 × 100 kW PDM	RIZ	1990
Ahwaz	Gulf	48.40E × 31.20N	Station not in service		

Of these SW sites the newest is at Sirjan which is equipped with ten 500 kW Telefunken S4005 transmitters, together with 48 curtain arrays and one rotatable curtain array. This turnkey project was completed in mid-1995 and was one of the largest SW projects awarded during the 1980s. It also included the supply and installation of six fixed curtain arrays and one rotatable curtain at the Kamalabad site.

Kamalabad is equipped with twelve 500 kW SW transmitters supplied by ABB between 1986 to 1988, two 250 kW and a number of 100 kW, also of ABB manufacture. Mashhad has four 500 kW transmitters and two 250 kW supplied by ABB between 1988 and 1993. Thus with twenty-six 500 kW transmitters (ten at Sirjan, twelve at Kamalabad and four at Mashhad), four 250 kW and a larger quantity of 100 kW transmitters, Islamic Republic of Iran Broadcasting has a total capacity on SW alone of at least 16 megawatts of carrier power, and the SW station at Kamalabad alone is one of the largest in the Middle East.

With twenty-six 500 kW SW transmitters alone, IRIB SW capacity exceeds that of all international broadcasters with the exception of Voice of America. Its external services are broadcast in the following languages: Arabic, Armenian, Assyrian, Azeri, Baluchi, Bengali, Chinese, Dari, English, French, German, Hausa, Italian, Urdish, Malay, Persian, Pushtu, Russian Serbo-Croat, Spanish, Swahili, Tajik, Turkish, Turkmen, Urdau and Uzbek. Foreign broadcasts are preceded by the announcement 'This is Tehran, the Voice of the Islamic Republic of Iran'.

Chapter 19
China and SE Asia

Since a lot of the time and energy of the US State Department is given over to China and certain other countries in SE Asia, it is appropriate that these countries should be given a broad geopolitical overview in this book. China, Cambodia, North Korea, Laos and Vietnam, according to the US State Department, all represent a threat to democracy and stability in the Far East and all feature as worthy targets in the Report of the President's Task Force on US Government international broadcasting.

China has the largest population on earth, 1.2 billion, and is generally accepted as the world's second superpower. With 30 different regions it requires a huge radio broadcasting infrastructure to cover national and regional broadcasting. Little is known about the infrastructure except that it has a strong dependency on AM broadcasting in the MW and SW bands. All radio broadcasting is controlled by the government through the Ministry of Radio, Film and Television. This same authority also operates special radio broadcasts to Taiwan and Vietnam. Broadcasts to Taiwan are identified by the name 'Voice of the Strait' and are managed by the People's Liberation Army of China. Another special programme for Taiwan is 'Voice of Pujian'. These are broadcast on MW and SW.

China operates a foreign service on SW under the name China Radio International (CRI). According to published data China is using a number of SW transmitter sites, at Baoding, Beijing, Jinhua, Kunming, Shijiazhuang and Xi'an. Quantities, types and powers of the transmitters are not published but are known to include transmitters with carrier powers from 50 to 500 kW. CRI has relay exchange agreements with Canada, Spain, France, Guiana, Russia and Switzerland. The US has never sought such an agreement.

According to programme output hours published by the BBC International Broadcast Audience Research (IBAR) unit for 1996, CRI was broadcasting 1620 hours per week, the second highest output in the world after the US with 1921. China has been buying substantial quantities of transmitters for MW, SW and for FM broadcasting.

155

China as an export market

In June 1995 the Harris Broadcast Division, based in Quincy, Illinois, USA, announced it had been awarded a contract to provide seven high power 600 kW MW transmitters for the People's Republic of China. The contract was awarded by China's Ministry of Radio, Film and Television. These Harris DX 600 MW transmitters replaced old tube type transmitters at transmitter sites within China. The contract included on-site commissioning and final acceptance for the first two transmitters, together with factory training at Harris Broadcast Division headquarters at Quincy. Negotiations began in 1994 and that August a delegation from the Ministry of Film, Radio and Television visited Quincy, to observe a Harris DX 600, the world's first all solid state 600 kW carrier power MW transmitter in operation. The delegation was evidently impressed by the solid state technology, believing it to be the future, and invited Harris to submit a formal proposal to the PRC.

This order for seven 600 kW all solid state transmitters comprised the highest powered transmitter to be supplied to China, but it is not the first major contract award from China, because in September 1994 the Dallas-based Continental Electronics Corporation (CEC) announced that it had secured a series of contracts for FM and SW transmitters, to be produced in the CEC Dallas plant and also in factories in Beijing. The deal between Continental and the Ministry of Film, Radio and Television significantly included extensive logistics such as technical training for the Chinese technicians. Looking ahead, according to Ross Faulkner, Vice President of Marketing at Continental, the Chinese Government estimates that it will require 50 000 FM broadcasting stations by 2006.

As the People's Republic of China enters the twenty-first century there will be a tremendous demand not only for radio broadcasting and television but for other services such as basic telephony, Internet access, business-to-business communications, cable TV and satellite TV on a direct-to-home basis. China has a stable economy with a GDP growth rate the envy of the Western world; moreover it is politically stable despite attempts from the West to move the philosophy of China to Western-style capitalism. China demonstrates with deeds an ability to pay for goods and other services from the West and its combination of a skilled labour force with a low wage structure makes China an attractive manufacturing base. Orders from China such as those awarded to Continental and Harris are the beginning of a huge potential transmitter market, developing on the following lines:

- Phase 1: Supply of complete transmitters in small quantities;
- Phase 2: Supply of components and sub-units to be assembled in Chinese production plants under guidance of Western technicians;
- Phase 3: Supply of certain components only, with everything else built in China;

- Phase 4: China reaches maturity in component mass production and becomes independent of Western technology.

Phases 1 and 2 are already in progress but phases 3 and 4 are a long way off and will gather momentum at a pace to suit China. China needs understanding because its culture and attitude to technology are fundamentally different to those of Japan, for example. Whereas Japanese technology is adaptive, Chinese is creative. Japan has taken well established technologies from the West, and has then honed and refined these products to a standard where they exceed those of the West and it has selected those products which are suitable for large scale mass production. Coupled with its adaptive skills Japan has developed a production philosophy and capability that has few equals in the Western world.

Chinese culture, by contrast, is dominated by original thought and may therefore be less inclined to follow the path taken by Japan. For several hundreds of years China has survived without the aid of any other nation. In more recent times it surprised the West with its first hydrogen bomb in 1957 and its highly successful rockets which have captured valuable launch contracts from the West. These give an indication of what China is capable of, and what the future holds for the largest nation in the world.

China is a paradox and an enigma. In some areas it leads the West, while in others, such as the production of modern consumer goods and professional equipments such as modern radio transmitters and television transmitters, it is behind the West. It is in such areas that China needs an injection of Western skills and production technology.

Back in the 1950s, and through to the late 1980s, the West remained highly sceptical about the People's Republic and the effect its brand of communism was having on other countries in SE Asia. Today that scepticism is being tempered by commercial awareness of the potential markets that are now opening up in China. Until its recent economic setbacks, SE Asia was the most buoyant and the fastest developing region of the world, and China (little affected by the economic crisis), is at the centre of this economic revolution. There is now a race by Western high technology manufacturing companies to climb aboard the Chinese band wagon.

Few experts would care to speculate just how long the business bonanza in China will last before it reaches the stage where China will be able to throw off the Western manufacturing companies, but most are agreed that one day it will happen. A Russian diplomat once observed 'They will cast you off as they cast us off'. It is encouraging to note that there seems to be no evidence of this happening yet.

The economic reality of China today with a stable economy, commitment to quality equipment and low employment costs, makes this vast republic the most attractive region of the world for joint ventures in technology transfer, with Western companies supplying the technology and China supplying production labour and management.

Korea

Korea has been split between North and South for the past 50 years, but it was not always so. There was a single Korean nation for several thousands of years, and even under Japanese occupation in World War II it was still one nation. The present-day tragedy is that not only are there two Koreas, but there are two cultures. One is the 'true' Korea perhaps, the Democratic People's Republic of Korea, otherwise known as North Korea. The other, South Korea, is a nation which has evolved with a massive injection of US culture and values, now in evidence throughout the country.

The Democratic People's Republic of Korea (DPRK) has a population of some 22 million, while the Korean Republic (i.e. South Korea) has a population of more than twice that number. US State Department claims that the citizens in the south live under a constant threat of invasion help to justify the huge presence of American forces in South Korea.

Broadcasting infrastructures

In the DPRK all radio and television comes under state control and this responsibility is vested in the Radio and Television Committee. The radio infrastructure is uncluttered and simple. National and regional broadcasting is done by the Korean Central Broadcasting station at Pangsong. It has three 250 kW, two 500 kW and one 1500 kW MW transmitters, as well as a number of low powered SW stations.

At Pyongyang Pangsong there is a second broadcasting centre with six MW transmitters. KRT (Korean Radio and Television), established on 14th October 1945, also operates a number of FM stations with powers from 1–20 kW.

Foreign service external broadcasting had a high priority right from the start and this got under way on 16th March 1947, initially in only the Chinese language. As the service expanded other language services were added; Japanese in 1950, English (1951), Russian and French (1963), Spanish (1965), Arabic (1970) and German (1983). External Services uses three main SW sites: Kanye with five 200 kW transmitters, Jujang also with five 200 kW transmitters and Pyongyang with ten 200 kW transmitters. Though KRT does not publish its programme hours per week, the BBC IBAR listed the figure for 1994 as 529 total direct programme hours, in seventh place in the world table. The corresponding figure for June 1996 was 364 programme hours, in 14th place in the table of broadcasters.

Republic of Korea

South Korea has a multiplicity of private and government broadcasters, far too numerous to list here. Of these, the biggest is the public service broadcaster the Korean Broadcasting System (KBS). The foreign service broadcaster is Radio Korea International (RKI), which operates two transmitter sites at Kimje and Hwasong. Kimje houses three 100 and three 250 kW transmitters, while Hwasong has two 100 kW transmitters. RKI has exchange relay agreements with the BBC World Service and Radio Canada International. Another foreign service broadcaster in South Korea is FEBC, a religious broadcaster, which operates a 100 kW MW transmitter to broadcast to North Korea and China on 1188 kHz.

Vietnam

Vietnam (population 74 million) is shaping up to become a very powerful player in SE Asian broadcasting, judging by its recent expenditure on high power MW transmitters. All Vietnamese radio broadcasting, both national and international, comes under the government-controlled Voice of Vietnam (VOV), the national broadcaster. It operates three networks for national, regional and provincial broadcasting. There are also some FM stations for serving urban centres, but in general VHF is not ideal because of the topography of the country.

VOV uses a number of SW sites and broadcasts in the following languages: Cambodian, Cantonese, Chinese, English, French, Indonesian, Japanese, Laotian, Russian, Spanish, Thai and Vietnamese. Geographically Vietnam borders China, Cambodia and Thailand to the east and Laos and Myanmar to the north, all within easy reach of MW – which might explain why Vietnam has made massive investments in high power and superpower MW transmitters. In August 1996 Harris was able to announce it had won the contract to supply VOV with one 2000 kW and three 500 kW transmitters, all for MW broadcasting. No other country in SE Asia has such powerful installations. The 2000 kW installation will deliver a strong signal over a wide region of SE Asia, possibly extending to the northern part of Australia. The use of MW transmitters on such a scale reinforces the fact that MW is now becoming very popular in Asia for international broadcasting.

Laos and Cambodia

These two Asian countries featured prominently in the US State Department Report from the US President's Task Force on US Government international broadcasting, published in 1991, which identified them both as targets for US government broadcasting on the grounds that they represented a part of the world where communist totalitarianism is still alive, 'the last redoubts of a pernicious ideology'. Laos borders Thailand to the west, Cambodia to the south and Vietnam to the east, while to the north its border touches Myanmar and China. Strategically and politically it separates Vietnam from Thailand, over a border which extends for more than 1000 km. During the Vietnam war it suffered from the invasion of the Ho Chi Minh army during its downward thrust into South Vietnam, and was also invaded by US troops with the support of South Korean forces.

Cambodia is a much smaller country in terms of area, situated at the southern end of Vietnam and Thailand. Both countries are very poor by Western and even Asian standards. Yet, from the American standpoint, so long as these two countries embrace communism they represent a threat to the survival of capitalism and US-style democracy in the Indo-China region.

Laos People's Republic

Laos has a population of approximately 4.7 million. It operates some fifteen AM radio stations of low to medium power except for one ageing 50 kW transmitter. There are some 575 000 radio sets in use and about 80 000 TV receivers. These low figures for radio and TV reception give us a good guide to the poor economic state of the nation. Laos National Radio and Television is the government body responsible for all national radio and television broadcasting and also its foreign service on SW.

This foreign service comes from one 50 kW SW transmitter situated at Vientiane. Its foreign service broadcasts go out on two frequencies, in the following languages at different time slots: Cambodian, 1 hour; English, 1 hour; French, 1.5 hours; Thai, 1.5 hours; Vietnamese, 1.5 hours.

Cambodia

Cambodia has a population of 9.7 million, 1.5 million radios in use and 70,000 TV sets. All radio and TV broadcasting comes under the control of National Radio and Television of Cambodia, which operates 4 MW radio stations, two of which have output powers of 120 and 150 kW. It also has a foreign service station at Phnom Penh which has one 15 kW and one 50 kW SW transmitter. Both are ancient, thought to be of Thomson-CSF manufacture and the fact that at least one is more than 50 years old is a tribute to the manufacturer. As with Laos, Cambodia broadcasts its foreign service in English, French, Thai and Vietnamese. For incoming foreign broadcasting, Radio France International programmes to Cambodia are re-broadcast over an FM station in the capital city Phnom Penh.

Myanmar

Myanmar, formerly known as Burma, was once a province of India. Situated on the Indo-Chinese peninsula, it adjoins Tibet, Laos and Thailand. It has a coastline of approximately 1900 km and its greatest breadth is 900 km. Much of the interior is a labyrinth of mountains which rise to 4000 m. Myanmar is one of the poorest countries in the world; its economy has deteriorated to the UN status of a least developed country (LDC). Only one in fourteen of its 14.4 million population owns a radio set. Its broadcasting infrastructure consists of an ageing 200 kW transmitter for national broadcasting on the medium wavebands and three 50 kW SW transmitters, two of which are more than 35 years old. These SW transmitters, all located at Yangon, are used for domestic broadcasting.

If it should seem disproportionate to devote a large part of this chapter to such an impoverished state of Asia, which has one of the lowest figures for GDP in the world, then it seems reasonable to point out that it is in keeping with the attention now being focused on Myanmar by the British Government and the BBC World Service.

During World War II a group of Burmese nationalists with little knowledge of the outside world were supplied with weapons by the British to fight the Japanese. As the war came to an end they formed the Anti-Fascist Peoples Freedom League (AFPFL), with a long-term objective of securing independence from Britain. The British Government was opposed to this and seriously considered the arrest of the leader of the AFPFL Aung San. He, along with other AFPFL leaders, was mysteriously assassinated in July 1947. However, their successors did succeed in getting the AFPFL treated as an embryonic government of Burma and they finally achieved the goal of full independence on 4th June 1947. The British then left Burma, leaving the country to solve its own internal problems which related to the economy and to ethnic divisions resulting from a population made up of Muslims, Catholics, Kachins, Arakans, Shans and others. Paradoxically, fifty years later, the British Government is engaged in a small cold war with the present military regime and is actively involved in trying to bring Aung San's daughter, Aung San Suu Kyi, to power following her election and the subsequent annulment of the result by the military.

The British Government is involved in attempting to overturn the present regime could be based on a fear that the 'anti-British' attitude of the present regime (at least perceived by the UK) could overspill to India and elsewhere in Asia. The problem is best seen in the context of the country's close neighbour, China. Since the 1950s Burma (Myanmar) has been haunted by a fear that if the US, with the assistance of Taiwan, launched a full-scale war against the People's Republic of China, then Burma would turn into a battlefield and suffer an even worse fate than Korea or Vietnam when the US went to war against them. Successive administrations have concluded that the best way of averting such an event would be to have independence from American and British presence.

Meanwhile, with a worsening economy, unrest from the poor and middle classes and discontent in the army, demonstrations have been brutally repressed, and the administration has changed hands several times. In 1989 the country's name was changed to Myanmar and at the same time the names of some towns and cities which until 1947 had played roles in the British Empire were changed – Rangoon became Yangon and Irrawaddy became Ayeyarwady, to name but two. Mandalay, which figured strongly in the days of the British Raj, appears so far to have kept its name. The country's subjugation by the British first began in March 1824 when Britain went to war with Burma and annexed Upper Burma. In April 1852 British forces were again used to annex Lower Burma. Finally in October 1885 British forces once again waged war and the British Government made Burma a province of the Indian Empire. The occupation was complete.

Britain's relationship with Burma since the 1970s deteriorated as one military dictator after another came and went, usually to be replaced with even harsher regimes, and relations reached an all-time low in 1987. This was the time when general Saw Maung dissolved parliament and promised new elections. Aung San's daughter returned from exile in England, won the elections, which were annulled, was arrested and subsequently released to degrees of house arrest.

Since 1988 the BBC World Service has reinforced its broadcasts to Myanmar, and on 21st August 1995 it reported that its broadcasts in the Burmese language were being jammed for the first time in the fifty-five years of history. BBC engineers found deliberate interference on two of the three regular frequencies in use on its Burmese language broadcasts. According to Marcia Poole, head of the Burma service, 'We do not fully understand why they are jamming us. We are not even sure it is Burma who are doing the jamming. It is true we released the news about the release of the opposition leader before that news had been given out by their own press'.

The BBC-WS should not have been surprised that the Myanmar government jammed its broadcasts, because they were obviously provocative. But the BBC-WS is no stranger to jamming – it has a record of being one of the most heavily jammed broadcasters and in nearly every case it claims to have been reporting impartial news and comment. A commonality with all international broadcasting where there is a political objective, and where relationships are strained, is to reveal to listeners in targeted countries the kind of news about internal affairs that would be suppressed by their own press and radio – this is called 'information broadcasting'.

Taiwan

For more than 40 years Taiwan has waged a war of words across the channel of the East China Sea that separates it from the eastern end of the mainland of China. With the assistance of America, Taiwan became the 'Republic of China'.

A vital part of the cold war between Taiwan and China is radio broadcasting. For a small country of 21.3 million, Taiwan has more broadcasting capacity than most countries in Asia. The main broadcasters are the Broadcasting Corporation of China (BCC), the Voice of Free China (VOFC), the Voice of Asia, the Central Broadcasting System and the National Association of Broadcasters (NAB). The last of these is a collection of public service and commercial radio stations. The rest are operated by the army or the government or are government-funded.

- BCC operates some 40 AM stations, mostly with 10 kW transmitters, and some of these stations target mainland China.
- VOFC is the foreign service broadcaster of BCC. It operates a number of SW stations, some of which relay programmes from the religious broadcaster WYFR (located in Florida).
- Voice of Asia is a government station which broadcasts on MW and SW. Its highest powered station is 1200 kW MW.
- Central Broadcasting System (CBS) is owned and operated by the Ministry of Defence. Easily the most powerful broadcaster, CBS operates a number of high power MW transmitters, some up to 1200 kW, and also broadcasts on SW.

For the most part, all the broadcasters transmit programmes in the Chinese language Mandarin, Taiwanese Amoy and in Hakka. VOFC in addition transits programmes in several other languages including Cantonese, English, French, German, Indonesian, Japanese, Spanish, Thai and Vietnamese.

The relay agreement referred to above between 'Voice of Free China' and the private US religious broadcaster WYFR is an example of how the privately owned international SW broadcasters in the US work in conjunction with foreign governments and their external broadcasters. In every case these foreign governments are right wing, they are mostly in Asia and all are allies of the USA.

The US and SE Asia

The most disputed post-war arena in SE Asia was Indo-China. The Khmer forebears of modern Cambodia ruled an empire which at its peak in the 12th Century contained parts of Burma, Siam, Laos and Annan. The French established themselves in this part of the world later than other European powers, although following the pattern of others by entering the competition for whatever as to be had out of the disintegration of the Asian empires. This, then, was the Indo-China empire over which the French

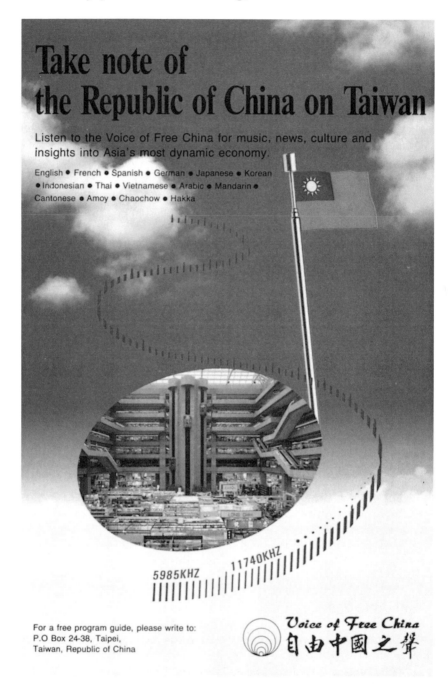

Figure 19.1 *Advertisement in WRTH for Voice of Free China, Taiwan*

ruled until World War II when the area was occupied by the Japanese. After the defeat of Japan it came under the control of the communist leader Ho Chi Minh. The scene was set for the French Vietminh war which was to last seven and a half years, and war lasted even longer when the US entered onto the scene. In the end both the French and the Americans suffered humiliation and defeat.

America's involvement in Indo-China in the 1960s was not its first venture into Asian politics. In 1945 when the US brought down the Japanese empire, and almost overnight found itself the legatee of a fallen empire that has stretched over the whole of SE Asia, it made political and strategic decisions that were to have a far-reaching effect on its national and foreign policy. The US supported the right wing armies of the KMT in an endeavour to halt the march of communist forces, and when Chiang Kai Shek's army fled to Taiwan, it continued to support this army in exile, pouring vast quantities of military aid into Taiwan. Not surprisingly, then, the newly established People's Republic of China saw the US as the arch-capitalist imperialist power and American values were discredited. Capitalism, rather than democracy, was the hallmark of US neo-colonialists; it oppressed China whilst at the same time building up the KMT forces in Taiwan and encouraged this force in exile to style itself as the Republic of China. It opposed the re-union of Korea under the communist regime and went to war in Vietnam. From the 1960s the US proceeded to put in place a structure of military bases which included Guam, Taiwan, Okinawa and the Philippines. Not surprisingly, smaller countries which had borders with China became nervous and some began to distance themselves from America and its allies fearing they might be engulfed in a bloody war.

War in Korea and defeat in Vietnam did not reduce US resolve against communism in SE Asia, and since the fall of communism in Eastern Europe and the USSR the US has turned its attention to other ways to meet its foreign policy goals in China and other countries of SE Asia.

What has happened since the end of the cold war has been a re-appraisal of the situation, with the idea of using the tool which assisted in the collapse of the Soviet empire – US surrogate broadcasting by RFE/RL – to bring about a a similar change in China and elsewhere in SE Asia. The Task Force document prepared for the US President in December 1991 laid down a strategy for accomplishing the elimination of communism. The following are extracts from that document:

- Five of the six remaining communist countries in the world are in Asia;
- When these countries change the world will be free of communism;
- To institute a new US surrogate radio broadcasting to the peoples of Asia living under totalitarian regimes;
- A substantial number of the Task Force agrees on the desirability of establishing a Radio Free Asia, to carry out surrogate broadcasting to

the communist countries of China, Vietnam, North Korea, Laos and
Cambodia. Such a service should start with minimum delay;
• These might be facilities of private US broadcasters, VOA transmitters,
or transmitters that belong to foreign nations.

Radio Free Asia was established in March 1996. Unlike RFE/RL during
the cold War era, RFA does not have dedicated transmission facilities and
its programmes are transmitted over VOA transmitters and on religious
broadcasting stations acting as surrogate broadcasters. The latest available
information from Washington suggests there are no plans for RFA to have
its own transmission facilities.

US options for increasing surrogate broadcasting capability

Strategies that might be deployed to strengthen the US Government's capability in surrogate broadcasting to China and the communist countries in
SE Asia are to take further leases on jammer transmitters in the former
Soviet Union or friendly republics in East and SE Asia, or use the US international religious broadcasters. The major criteria in the choice of SW
transmitter facilities are political considerations, the cost of leasing, and the
transmitter power available, signal coverage quality and distance from the
target zone. Sites should ideally be about 2500 miles (4000 km) distant.
Being too close to the target zone can produce a lower quality of signal in
terms of signal strength.

VOA is already leasing air-time on some Russian jammer transmitters
in Siberia. Technically, this is an attractive option, but the Russians are
increasingly seeking higher payments. As for the friendly-disposed republics
in SE Asia, such countries may not want to become enmeshed in what could
turn out to be 'Cold War II' – between the US and the People's Republic of
China. On balance the most attractive option seems to be to enlist the aid of
US privateers; the US international/religious broadcasters. There is a substantial number of these broadcasters, some of which have their SW transmitter sites in North America, though the vast majority possess SW
transmitter facilities in the Pacific islands and the Indian Ocean (see
Chapter 21).

The advantages of enlisting the services of these religious privateers are
twofold. Some might be persuaded to hire out a complete SW facility, or
perhaps even build an additional SW transmitter on their transmitter
station, for US surrogate broadcasting. However, most would also be
receptive to giving air-time to the government broadcasters in times of
international political crisis, when the US government international broadcasting operations might need 'surge capacity', adding more transmitters
on-air in order to get the news through to a specific country or target zone.
Such examples include the time of the Gulf War and, before that, the

Tiananmen Square riots in China in June 1989. In fact most of the religious broadcasters have an option in their structure of programming operations which permits them to broadcast programmes for other organisations. The US Task Force on International Broadcasting indeed reported that there were good reasons to believe that some of the religious broadcasters' facilities might be made available to US Governmental broadcast interests into China and elsewhere.

Conclusions

The inherent dangers associated with waging a cold war with words and not bullets are discussed in *History of international broadcasting* [2]. In the context of that book, the war was with the Soviet Union which was hostile to the ambitions of the US to become the world's only superpower, and for that reason (and a few others) the political relationship between the US and the Soviet Union was always at a dangerous level.

This is certainly not true with China. It has not sealed itself in a vacuum, it has open borders, it trades with the world's countries irrespective of whether they are a right wing capitalist, communist or revolutionary regime. The strained relations that exist between America and the People's Republic are mostly one-sided and seem to stem from an ingrained US belief that the world will be a safer place only when communism has been eradicated. Towards this goal the United States projects American-style democracy, beliefs, culture and the American way of life to the world at large.

It is this author's contention that the direction of broadcasts over the heads of the established leaders, or governments, of the few remaining communist countries in the world carries an element of danger, which though it might be small at the onset, has the potential to escalate just like it did in the cold war with the USSR.

The cold war between the West and the Soviet Union was the first time any nation had ever sought to defeat another with words and not armaments. Yet history has shown that it was a high-risk strategy. Barrage broadcasting on all available SW frequencies by Western broadcasters, and unparalleled investment by the Russians into hundreds of sophisticated jamming installations, produced a state of 'oscillatory antagonism'.

Propaganda broadcasting will always find an audience. During the cold war broadcasts from Radio Moscow found sympathetic listeners in the United Kingdom and elsewhere in Western Europe. These came from different strata of society (the working classes, academics and from the literary world) but were not in sufficient numbers to constitute a threat to government. In the case of the People's Republic of China, the US administration believes listeners exist in sizeable numbers. If this prognosis is correct, then US surrogate broadcasting to China would encourage small groups seeking political change, possibly even to stage uprisings on a bigger scale than the Tiananmen Square riots in June 1989, which China believes were partly initiated by VOA and BBC World Service broadcasts.

Radio Free Asia and Radio Free China: estimated set-up costs

In 1991 RFE/RL engineers were requested to provide estimates for a fully-fledged surrogate broadcasting service to China, Vietnam, Cambodia, Laos and North Korea. RFE/RL engineers saw Radio Free Asia and Radio Free China operating in much the same way that RFE/RL did from its Munich studios, with a round-the-clock broadcasting service. RFE/RL proved broadcasting in the early hours was cost-effective during the cold war, and though the numbers of listeners through the night may be few in number, this audience was often composed of hard core listeners, many of whom were dissidents.

Similarly, effective surrogate broadcasting calls for intelligence and news-gathering operations on what is happening inside the targeted countries. To set up a Radio Free Asia and Radio Free China service, RFE/RL estimated a need for 60 programmers, 25 news-staff and an absolute minimum of 20 additional analysts, archivists and librarians. The final vital requirement was that of monitoring. This would need listener–agents in the various countries who would report back on effectiveness. Adding in transmitter staffing and general administrative personnel would bring the total for staffing to 200. In terms of costs, RFE/RL engineers estimated that an efficient surrogate broadcasting operation for China could be delivered for $90 million, based on six 500 kW SW transmitters and antennas. Additional start-up costs would be $19 million and the annual running costs would be $34 million. A further six 50 kW SW transmitters serving Radio Free Asia could be installed at the same transmitter site.

Chapter 20
Radio Australia

Radio Australia's role in World War II

Almost from the start of World War II, German broadcasting in the SW bands was on a war footing, broadcasting to the world in several languages with two objectives: to justify its position with neutral nations and to weaken the morale of Allied forces. At this time Australia's part in the counter-propaganda war was small. However, on the entry of Japan on the side of Germany the situation changed. The Pacific had become a war zone where Japanese broadcasting in the SW bands became a dominant player. As the propaganda war intensified, with both Japan and Germany playing powerful roles, it was time for Radio Australia to play a more substantial part [2].

Radio Australia had actually joined the war effort in December 1940 but suffered from the fact that its available transmitter power was no match for the superior power of the Axis countries, either in the transmitter power or in the number of SW transmitters. The Australian counter-propaganda service started using whatever transmitter resources could be found. A limited service commenced using an Amalgamated Wireless Australia (AWA) transmitter located in Sydney, and this was supplemented by its own stations VLW and VLR whenever practical. The first-mentioned transmitter became available owing to the suspension of certain radio-telephony international circuits. AWA had a nominal carrier power of 10 kW and operated with call letters VLQ. VLR and VLW had a carrier power of 2 kW each, which meant that the total transmitter power of Radio Australia at the beginning of the war was just 14 kW. However, a surprisingly good job was done despite this, in the face of total German transmitter output power of many hundreds of kilowatts. It should be remembered that during the 1930s and 1940s the SW bands were uncluttered and man-made noise was low; in consequence even a low powered transmission on SW could be clearly audible over a distance of several thousand miles.

However, the need for Australia to have a high power SW capability was evident. Accordingly, a proposal to establish a powerful high frequency transmitting centre in Australia was discussed in 1941 between the British and Australian governments, as a result of which the Postmaster General's engineering department drew up a plan that ultimately became Shepparton. This proposal was none too soon because by this time (October 1941) Japan had entered the war.

A number of factors contribute to the ideal transmitter site. These include propagation requirements, availability of power and other services, topographical requirements (a flat landscape), soil conductivity, water table and access and proximity to the programme centres. Finally, from a short-list of suitable sites, the PMG engineering department settled upon Shepparton in Victoria, a site that was flat for long distances in every direction. The area of the antenna site was 567 acres.

By the end of 1942 the sheer speed of the Japanese onslaught upon both Allied and neutral territories was having a dramatic effect. After the rapid collapse of Singapore and Malaya, Japanese forces quickly moved through the Dutch East Indies, Java, the Solomon Islands and down to New Guinea, and in the process quickly captured many radio stations and communications facilities. The 14 kW of power that the Australian Broadcasting Corporation (ABC) had at its disposal for SW broadcasting was no match for the power of Japanese propaganda over the short waves. Indeed, Japanese propaganda broadcasts to Australian forces in the Pacific usually referred to ABC as 'the BBC's little sister'.

The Shepparton SW complex was completed in 1944, and in May that year it came on-air, equipped with two 100 kW SW transmitters manufactured jointly by STC (Standard Telephone & Cable) and AWA, supplemented by a 50 kW SW transmitter supplied by the RCA Company.

The transformation of ABC's foreign service on the short waves, from a poorly equipped service in the 1930s, with just a few kilowatts of transmitter power, to a world broadcaster by the mid 1940s is one more proof that politics and wars act as a spur to technology. By 1945 the Shepparton HF centre – by now officially christened Radio Australia – had begun to broadcast news and entertainment to all parts of the world in addition to its prime role broadcasting to the Allied forces in the Pacific. Its broadcasts were regularly picked up and relayed over other broadcasting networks and Allied networks. These included the BBC, the South African Broadcasting Corporation and some in the US. Inward mail figures to Radio Australia showed that its biggest listening audience was in Britain.

The RCA transmitter used at Shepparton represented the state-of-the-art in the early 1940s [10]. It delivered 50 kW carrier power over the range 6–22 MHz, and comprised two complete radio frequency channels from crystal to output and one Class B modulator which could modulate either RF unit at high power. Each channel comprised six states, the first being a crystal oscillator using an 802 tube with six switched crystals on different frequencies. This stage fed an 802 tube in a doubler stage which in turn fed a pair of 828 pentodes which required no neutralisation. Following this were a pair of 810 triodes using cross-condenser neutralisation. These drove the penultimate stage with two 827R external anode, airblast-cooled tetrodes. The final output stage using water-cooled type 880 triodes, cross-condenser neutralised. The inductors of this stage were unusual in that they employed long copper tubes with adjustable shorting bars for coarse tuning. No less complicated was the design of the modulator, this used a push–pull arrangement of Class A amplifiers which was driven by four previous stages of audio amplification.

Radio Australia: the present

Radio Australia is the overseas service of the Australian Broadcasting Corporation. As with almost every other foreign service broadcaster, Radio Australia is public funded through appropriations from the government. Radio Australia operates four SW stations at Shepparton, Victoria, Carnarvon in Western Australia, Darwin in the Northern Territories, and Brandon in Queensland. At the end of 1996 the numbers of SW transmitters in service were as shown in Table 20.1 with a total transmitter power of 2520 kilowatts. The latest acquisition has been at the Darwin site, which was fitted with two 250 kW type TRE2326 Thomson CSF transmitters in 1993.

Table 20.1 *Radio Australia SW transmitters, 1996*

Site	Coordinates	SW transmitters	Manufacturers
Shepparton	6 × 100 kW	145.25E × 36.20S	Harris
Carnarvon	1 × 300 kW 1 × 250 kW 1 × 100 kW	113.43E × 24.54S	Thomson-CSF Thomson-CSF Harris
Darwin	2 × 250 kW 3 × 250 kW	130.38E × 12.25S	Thomson-CSF CEC
Brandon	2 × 10 kW	147.20E × 19.30S	STC

Radio Australia broadcasts from its Melbourne studios in nine languages: Cantonese, Chinese, English, French, Indonesian, Khmer, Thai, Tok Pisin and Vietnamese. Its SW broadcasts are easily identified by its 'Waltzing Matilda' which is played for a few minutes before opening up on all frequencies, and all foreign language broadcasts start with the laugh of the kookaburra bird. This has been the hallmark of Australian broadcasting on the international SW bands for more than sixty years. Radio Australia's operating budget for 1995/6 was Aus$30 million and in the opinion of many other international broadcasters it does a fine job with a budget which is by no means excessive when compared with some other countries. Yet in the economic climate of today, where there are many demands upon the funding of public service broadcasting of radio and television, the role of SW broadcasting is coming under increasing scrutiny as to whether the cost is justifiable, and even this modest budget may have to be reduced.

Radio Australia may be a relatively small player when compared to the likes of Voice of America or the BBC World Service. Nevertheless it is one of the most respected broadcasters on the SW bands. It is one of the small band of pioneering countries, along with Britain, Germany, the Netherlands, Russia and a few others, which began SW broadcasting between the late 1920s and the early 1930s. In the case of Australia the forces which propelled it to become a pioneer of that art and science were the urge to open up communications with its 'mother' country and also the huge internal distances that had to be spanned, so its transmission engineers became skilled in the art of multi-hop via the ionosphere layers, required to span the 12 000 miles or so to Europe.

According to information published by the International Broadcasting Audience Research (IBAR) Radio Australia's weekly output was 307 hours in 1995, a figure which is some reduction from its 1970 figure of 350.

Australia: internal SW services

Australia is one of the few countries in the world that has, from the early origins of SW, used it as a carrier for national broadcasting. Occupying a huge landmass somewhat greater than that of the US, but with a relatively small population including large tracts of desert, SW was the ideal medium because of its ability to span the country's breadth in a single hop. MW, on the other hand, has a limit of a few hundred miles at best, dependent on output power, operating frequency, soil conductivity, the general nature of the country and latitude. Thus MW is useful in closely settled areas of higher density population such as the eastern and south-eastern coastal regions and a few parts of south and west Australia. Although MW broadcasting has a sky-wave component after the hours of darkness, this is not always a satisfactory service due to fading. And as some parts of Australia are subject to tropical storms where a high level of atmospheric noise is a frequent occurrence MW is not ideal. SW, on the other hand, is less susceptible to atmospheric noise and the correct choice of frequency can help to offset fading.

Recognition of these factors led Australia towards using the short waves as early as 1928. That first service was inaugurated with VK3LR. It used a power of 500 watts and transmitted from the PMG research laboratory at Lyndhurst, near Melbourne. By 1939 the station had expanded to a 10 kW transmitter along with a few smaller transmitters. Today, nearly sixty years later, ABC operates the Northern Territories Shortwave Service with programmes in English and Aboriginal language. SBS (Special Broadcast Service) also operates a limited service on SW.

Chapter 21
US religious/commercial private broadcasters

Today in the US there are over 12 000 radio stations, around 12.5 per cent of which (nearly 1600) broadcast fifteen or more hours of religious pro-grammes per week (see Table 21.1). National religious broadcasting is one of the fastest growing sectors of broadcasting in America. These radio stations are generally found on FM and on AM MW, however there are a number of religious radio stations using SW. These international radio

Table 21.1 *Total religious radio stations in US, 1971–1997*

Year	Radio stations full-time	Radio stations part-time	Total
1971	399	NA	399
1974	640	NA	640
1975	531	840	1371
1976	765	295	1060
1978	946	264	1210
1979	905	57	962
1981	736	91	827
1983	869	125	994
1984	894	163	1057
1985	991	174	1165
1986	1146	294	1440
1987	962	353	1315
1988	1139	359	1498
1989	1177	365	1542
1990	1140	252	1392
1991	1156	258	1414
1993	1084	482	1566
1994	1328	NA	1328
1995	1463	NA	1463
1996	1648	NA	1648
1997	1240	292	1532

Source: 1998 Directory of Religious Media

stations broadcast to audiences throughout continental America but chiefly to Central American countries, the Caribbean and South America, where SW listening is popular. A SW station in the USA has to be licensed by the FCC (Federal Communication Commission), for which it has to satisfy certain conditions, one of which is that the applicant has to establish that there is a need for its programmes. Religion appeals to mass audiences particularly in the south and the mid-west of the US – the 'bible belt'. Topical issues can be made straightforward; good versus evil, God versus the devil. One-dimensional messages such as these can have strong appeal.

Table 21.2 shows a list of SW stations based in the 'Conus' (Continental US) region with their operator or company, location, number of transmitters and carrier output powers. This represented the position at the end of 1996, though new stations appear each year and, equally, some go out of business or the transmitting stations change hands. The majority of the stations are of religious denominations, but a few are purely commercial. Most are small private companies but there are a number of heavyweights. Monitor Radio (WSHB), for instance, has two 500 kW SW transmitters. It is the world service broadcaster of the First Church of Christ, Scientist in Boston, MA and this broadcaster is more of the genre of major international broadcasters, such as the BBC World Service, being a global rather than a regional broadcaster. WSHB, along with station KHBI located in Saipan, covers all regions of the world. Monitor Radio builds on the 86 year old journalistic tradition of the 'Christian Science Monitor' newspaper found in libraries throughout the world. All the news and religious programmes that are broadcast are prepared by Christian Science Monitor's own staff.

Table 21.3 is a summary of US religious broadcasters operating from SW facilities in the Pacific and Indian Ocean regions. Most are in commonwealth territories of the US, such as Guam, Saipan and Palau in the Micronesian Islands, whilst others are in the Seychelles and the Philippines. They are not required to be licensed by the FCC in the US, but come under the licensing conditions of the country concerned.

The strategic and technical advances of using island bases cannot be over-stated. Apart from the obvious advantage of being at optimum range from target areas in Asia, these coral and volcanic islands possess near perfect propagation characteristics. The ocean presents excellent fresnel zone characteristics with a high order of electrical conductivity. Thus, by positioning the antenna out beyond the shoreline at sea, erected on a coral reef as in the Seychelles, on a volcanic reef such as those found in the Northern Marianas or even on a man-made reef, the sea in the foreground will permit the use of very low angles of departure above the horizon down to a grazing angle, i.e. 2–3 degrees. This will ensure the maximum path distance for a single ionospheric propagation. For long distance paths it is desirable to use the lowest angle of departure because, besides achieving that objective, it also ensures minimum absorption in the reflective ionospheric layer, and therefore maximum audibility in the target area. For all these reasons, plus

Table 21.2 *US religious/commercial broadcasters with SW sites in the Conus region*

Broadcaster	Location	Rel/com	Tx site	Coordinates	Transmitters	Call sign
KAIJ International	Washington DC	Rel	Denton, Tx	96.52W × 33.13N	1 × 50 kW	KAIJ
Radio Station KJES	Mesquito, NM	Rel	Vado, NM	106.35W × 32.08N	1 × 50 kW	KJES
KTBN International	Santa Ana, CA	Rel	Salt Lake City, UT	112.03W × 40.39N	1 × 100 kW	KTBN
High Adventure	Van Nuys, CA	Rel	Rancho Simi, CA	118.38W × 34.15N	1 × 50 kW	KVOH
Worldwide Catholic	Birmingham, AL	Rel	Vandiver, AL	86.28W × 33.30N	4 × 500 kW	WEWN
World Harvest Radio	South Bend, IN	Rel	Noblesville, IN	86.57W × 40.01N	2 × 100 kW	WHRI
Assemblies of Yahweh	Bethel, PA	Rel	Bethel, PA	76.17W × 40.29N	1 × 50 kW	WMLK
WJCR Worldwide	Upton, KY	Rel	Millerstown, KY	86.02W × 37.26N	3 × 50 kW	WJCR
WYFR Family Radio	Oakland, CA	Rel	Okeechobee, FL	80.39W × 27.46N	9 × 100 kW	WYFR
WHVA	Mt. Dora, FL	Rel	Scotts Corner, ME	68.34W × 45.09N	1 × 500 kW	WHVA
Radio Miami Int'l	Miami, FL	Com	Miami, FL	80.21W × 25.54N	1 × 50 kW	WRMI
World Int'l Broadcasters	Red Lyon, PA	Com	Red Lyon, PA	76.34W × 39.54N	2 × 50 kW	WINB
WRNO Worldwide	Metairie, LA	Com	Marrero, LA	90.07W × 29.50N	1 × 100 kW	WRNO
Monitor Radio	Boston, MA	Com	Cypress Creek, SC	81.07W × 32.41N	2 × 500 kW	WSHB
Worldwide Christian Radio	Nashville, TN	Com	Nashville, TN	86.53W × 36.12N	4 × 100 kW	WWCR

Table 21.3 *US religious broadcasters in the Pacific and Indian Ocean regions*

Region	Island(s)	Broadcaster	Coordinates	Transmitters	Call sign
Pacific Ocean E.	Guam	AWR	144.39E × 13.20N	4 × 100 kW	KSDA
Pacific Ocean E.	Guam	TWR	154.40E × 13.17N	4 × 100 kW	KTWR
Pacific Ocean E.	Saipan	FEBC	145.50E × 15.16N	4 × 100 kW	KFBS
Pacific Ocean E.	Saipan	Christian Science Monitor	145.41E × 15.07N	2 × 100 kW	KHBI
Pacific Ocean E.	Palau	Voice of Hope	144.42E × 13.28N	2 × 100 kW	KHBN
Pacific Ocean W.	Hawaii	World Harvest Radio	155.40W × 19.01N	1 × 100 kW	KWHR
Indian Ocean	Seychelles	FEBA	55.28E × 04.36S	4 × 100 kW	
Pacific, Philippines	Bocaue	FEBC	120.55E × 14.48N	1 × 50 kW	
Pacific, Philippines	Bocaue	FEBC	120.55E × 14.48N	1 × 100 kW	
Pacific, Philippines	Iba	FEBC	119.58E × 15.20N	1 × 100 kW	
Pacific, Philippines	Iba	FEBC	119.58E × 15.20N	1 × 100 kW	
Pacific, Philippines	Palauig	Radio Veritas	119.50E × 15.28N	3 × 250 kW	

AWR: Adventist World Radio (headquarters in California)
TWR: TransWorld Radio (headquarters in N. Carolina)
FEBC: Far East Broadcasting Co. (headquarters in Silver Spring, Maryland)
Voice of Hope (headquarters in California)
Christian Science Monitor (headquarters in Boston, MA)

their proximity to China, the US Commonwealth Pacific islands will continue to be the preferred choice for many of the international and the religious broadcasters.

Religious broadcasting has been called the 'Electronic Church'. The notion of using the short waves to project a faith had its origins in the 1930s but it is the past two decades that have seen the most dramatic growth. The logic of the religious broadcasters is that such a tool can largely replace the missionary in developing countries. Adventist World Radio (AWR) has estimated that a SW facility with two transmitters and a half-dozen directive curtain arrays can accomplish what it could have taken 10 000 missionaries to do in a sub-continent the size of India. Moreover, with the aid of pre-prepared cassette tapes in as many as 100 different languages and dialects, the religious broadcasting station reinforce the message of the missionary far more efficiently.

Due to the nature and type of programming, the costs associated with the running of a religious broadcasting station are consistently lower than those of an international broadcaster. Programme production costs are lower and many of the staff are believers, and work in donated service for little or no payment. AWR, one of the biggest of the religious broadcasting companies, estimates its annual operating costs at $5 million. The Far East Broadcasting Company (FEBC) estimates its annual budget at $17 million. The extraordinarily low operating costs of AWR are helped by the fact that it uses its churches in overseas countries as studios in many cases, and the operating staff are drawn from these local churches.

Religious broadcasting companies claim that their operating costs are borne from donations and legacies, sometimes with a proportion coming from programme suppliers. Notwithstanding the apparently random and irregular source of income, the major religious broadcasting companies have long-term business strategies. For example, AWR began to plan for a major new SW transmitter facility in the late 1980s. A suitable site was purchased in the area of Argenta, in northern Italy, with the Adriatic Sea 25 km to the east, and tender documents were issued in September 1996 for construction of the facility.

This tender called for operational capability to deliver signals with high audibility to the specified prioritised mission areas of the Middle East, Western Europe, all of Africa, Central Asia and Indo-Asia, totalling over 100 target countries. The tender documents do not specify numbers or powers of transmitters or directive curtain arrays, but a system analysis indicates something like a total of six SW transmitters, two 250 kW and four 100 kW, including a spare, would be appropriate. Such a facility would cost of the order of $50 million with the antennas and all other ancillary items. That the international religious broadcasters budget such sums to project the faith to more than 100 countries worldwide gives some idea of the religious and political importance of such projects to the broadcaster.

Table 21.4 *Output in programme hours for the major religious broadcasters*

Broadcaster		Program hours per week
Adventist World Radio	(AWR)	806.50
TransWorld Radio	(TWR)	725.50
Far East Broadcasting Co.	(FEBC)	631.50 ⎫ FEBC Intl. 774.00
Far East Broadcasting Assn.	(FEBA)	142.50 ⎭
Voice of the Andes	(HCJB)	485.43
The Vatican		344.00
Veritas		127.50

Source: IBAR, BBC World Service, 1996

As powerful as these Christian missionary broadcasters are, they co-operate at high level to achieve common goals. In 1985 Paul Adams, President of TransWorld Radio (TWR), and Jim Bowman, President of FEBC International, got together with the president of World Missionary Fellowship, the operator of station HCJB La Vox de Los Andes, to determine common goals. The result of that meeting was the founding of WORLD 2000, a co-operation between all major Christian broadcasters with the aim of bringing the Christian gospel to every person on the planet, by 2000.

TWR, one of the three biggest religious broadcasters, targets Eastern Europe, the Near and Middle East, some of the former Soviet republics and parts of Africa. FEBC broadcasts to the Philippines, to certain countries in SE Asia, India and China, whilst AWR, the biggest of all, covers the five continents. Table 21.4 shows programme hours per week for the seven most powerful broadcasters.

Involvement of US religious broadcasters in transmissions to China

The involvement of US religious broadcasters with China began at the time America was supporting the KMT (Kuomintang) forces of Chiang Kai Shek in the 1940s. By one of those remarkable coincidences that seems to be a feature of US foreign policy, the Far East Broadcasting Company (FEBC) was formed in 1945 by Robert H Bowman and John C Broger for the purpose of broadcasting Christian radio to China. Refused a licence by the Chinese government, Broger visited the Philippines where in August 1946 FEBC (Philippines) Inc. was founded.

The first FEBC broadcast took place in 1948 and in 1949 it started its SW international service to China. According to FEBC published documents, FEBC's broadcasting network expanded to take in SW transmitter sites in the Northern Philippines, Okinawa, Korea and the Seychelles. The combination of this transmitter power constituted the inauguration of 'Open Door to China': Phase 1.

Phase 2 began in 1975 when a MW site was acquired at Iba, 120 miles north of Manila, where a 250 kW transmitter came into operation, followed by two 100 kW SW transmitters in 1986. At about this time FEBC had formed a UK-based associate company, Far East Broadcasting Associates (FEBA), and three powerful FEBA 100 kW SW transmitters went into service from a site in the Seychelles. Together with a further two FEBC SW transmitters operating from Saipan in the Northern Marianas, the FEBC religious broadcasting group was able to announce that 'Open Door to China': Phase 3 was inaugurated.

FEBC and its British-registered sister company FEBA are not the only Christian broadcasting association to target the People's Republic of China with SW broadcasts from strategically placed transmitter sites. Another is Adventist World Radio (AWR), the broadcasting arm of the Seventh Day Adventist Church which came into being in 1971, modestly at first but expanded rapidly from the mid 1980s. Today it operates in five main regions of the world – AWR Africa, AWR Asia, AWR Europe, AWR Middle East and AWR Pan America – and it has come to be regarded as one of the major players.

More details on the AWR is given later in this chapter, but the following concerns its broadcasts to China. Two Thomson-CSF 100 kW SW transmitters went into service on Guam on 7th March 1987, with four highly directive curtains targeted on mainland China. A third 100 kW SW transmitter, this time from Continental Electronics, was added in August 1994, and a fourth Continental transmitter went into service in January 1996. AWR makes no secret of the purpose of the Guam transmitter site. A published account of AWR operations states, 'Our Guam station's primary target is mainland China and results show that that target is being hit. In the late 1980s it was estimated that between 20 and 30 thousand people joined the church each year in China' [11].

The book [11] also says 'Within eighteen months of commencement of broadcasts to China Guam AWR received thousands of letters from all provinces of China' and also cites a news service report in 1990, 'In recent months students have been converting to Christianity in large numbers, the rate is literally a dormitory at a time.'

Another US international religious broadcaster located in the Pacific Islands is TransWorld Radio (TWR), the oldest of the privately owned religious broadcasters and amongst the most powerful. It operates four 100 kW SW transmitters at Marpi in the Northern Marianas. The first two went into service in 1984, a third was relocated from Redwood City, USA to the island of Okinawa and finally to Marpi in 1985, and a further two went on air the same year, when the oldest transmitter was taken out of service to leave four in service.

Thus mainland China is being targeted with the SW broadcasts of a religious nature from several sites to the east and south-east (Table 21.5). Broadcasts from US international religious broadcasters are in addition to those carried out to China by Voice of America, which the US Government's International Broadcasting Bureau (IBB) states will be expanded in the future. A new high power SW transmitting station for IBB is in the planning stage, to be located on the Pacific island of Tinian.

Table 21.5 *Religious broadcaster sites targeting China*

Location	Broadcaster	Transmitters number and power
Guam KSDA	AWR Asia	4 × 100 kW SW
Northern Marianas KFBS	FEBC	4 × 100 kW SW
Philippines	FEBC	3 × 100 kW SW
Cheju-do, Korea HLAZ	FEBC	1 × 250 kW MW
Inchon, Korea HLKX	FEBC	1 × 100 kW MW
Seychelles	FEBA	4 × 100 kW SW

International religious broadcasting and global politics

Over the past few decades the international religious broadcasters have come to be recognised by Western governments as a potentially powerful ally on the short waves. The advantage that these broadcasters bring is they are promoting faith in religion, the powerful unifier of people. The listeners may come from different walks of life, backgrounds and cultures, but they can be welded into a powerful army.

International religious broadcasters differ from international broadcasters such as Voice of America, whose main function is to disseminate world news on a a broad basis; the religious broadcaster aims to propagate beliefs and faith. This is one of the reasons why religious broadcasters yield the power they do, and why they are able to exert considerable influence in many regions of the world.

Before the end of World War II, the Allies were beginning to realise the size of the threat of Soviet communism in Eastern Europe, China and in East and SE Asia. Some were in the third world, some were of primitive culture and some were oppressed. All were thought to be susceptible to communist propaganda with its message of people power. This fear became fact when, after the war ended, the US found its post-war foreign policy threatened by the new founded People's Republic of China. Both America and Britain realised the possibility of the 'domino effect', i.e. more nations in Asia coming under the influence of Chinese communism.

In their search for a strategy to fight communism, Western powers realised that there was a commonality between religion and communism. Communism is a philosophy based upon a set of beliefs, in other words an ideology or faith, the same qualities found in any living religion. No proofs are supplied and those who question the doctrines can be branded as non-believers and threatened with dire consequences. In a contest of ideas between logic and a faith, reasoned logic will always be the loser because the propagator of the faith always seeks to place the onus of disproving the faith onto the non-believer. Thus it came to be realised that a faith can best be defeated by the substitution of another faith.

Christianity came to be perceived by the West as a powerful tool in its fight against communism. Religious propaganda is able to communicate in relatively simple terms, and in a way that poorly educated people can understand. Issues can be simple – either Christ or anti-Christ, good or evil, from which it is a simple extension to identify Christ with goodness, and the anti-Christ with evil or as the force of tyranny and oppression. One-dimensional messages such as these can have an instant appeal to the poor or oppressed. In the hands of skilled preachers, quotations from the Bible can have considerable potency. Religious propaganda can be used to unify a race of people but equally it can be used to disrupt a nation by the promotion of passive and even armed resistance to the government in power. Political leaders of many nations and most religions, from Christianity to Islam and the Jewish faith, have found it politically advantageous to bring God into the picture. When George Bush decided to go to war with Iraq he quoted in his opening speech at the 1990 US National Religious Broadcasters (NRB) convention a verse from Ecclesiastes, 'There's a time for peace and a time for war'. Likewise, when he was drumming up moral, political and military support from other Western nations he found another apt quotation 'My concern is not whether God is on our side, but whether we are on God's side'. It is a matter of record that Bush at the NRB convention in 1995 commended several religious broadcasters for their work in the Gulf War.

The first task of an international information broadcaster is to gain the trust of its listeners as the first step to building up a sizeable listening audience. Religion appeals to different people in different ways. To most it is usually seen as wholesome and stimulating. It can also give hope and succour, provide a warm and evangelical feeling – all with the aim of building up trust.

Stemming from trust in the minds of listeners, a station is able in its broadcasts to call for an action that seems to be reasonable. Moreover, the broadcaster can suggest that a certain course of action is right. Religious broadcasting stations are able to convey the feeling to the listener of his or her belonging to an elite cause. To emphasise this feeling of belonging, international SW stations seek to recruit listeners by such means as setting up SW listeners clubs and the issue of membership certificates, souvenirs and QSL cards.

Figure 21.1 *QSL card for KHBI station, Saipan, N. Marianas*

Figure 21.2 *QSL card for WSHB, Cypress Creek, SC, US*

An overview of Adventist World Radio (AWR)

Adventist World Radio is the international broadcasting arm of the General Conference of Seventh Day Adventists, that owns and operates all AWR facilities. AWR's headquarters are in Silver Spring, Maryland. The Seventh Day Adventists are a Christian movement established in 1844, initially part of the Adventists founded by William Miller in 1831. For over 100 years it has propagated its faith from its many churches around the world, so a move to radio was a logical extension.

From a small beginning in the early 1970s, AWR has grown to become one of the most powerful of all the international religious broadcasters. Its SW facility at Facti Point in Guam is one of the most powerful SW stations in the Pacific, with four 100 kW transmitters (see p. 177). In terms of programme hours per week, the BBC's IBAR unit registered a total 806.30 weekly broadcast hours by AWR for the month of January 1996, and since that audit AWR has continued to increase its programme output. Table 21.4 (p. 179) compared this figure with other major international religious broadcasters.

Where AWR operations differ from some of other religious international broadcasters is that it does not commission third parties to prepare its programme material. Instead, AWR prepares its broadcasts in its own studios, usually in the country to which the broadcasts are to be targeted. Not only can this method be more efficient in terms of production costs but it ensures a higher standard of accuracy in terms of syntax. In the case of China, its programmes are prepared in the AWR studio in Hong Kong and broadcast from Guam where AWR has its high power SW transmitter. AWR makes no secret of the fact that the People's Republic of China is one of its most important targets; it prepares tapes in the five most widely spoken languages (Mandarin, Cantonese, Fujian, Akka and Shanghai dialect).

AWR and the new Russia

It was a cold −20° C when AWR board chairman Kenneth Mittleider and president Walter Scragg threw the big switch that inaugurated the first ever AWR radio broadcast from Russia. The date was 1st March 1992, just two years after the end of the cold war, and Adventist World Radio was the first religious broadcaster to begin utilising Russian government-owned high power SW transmission facilities in the former Soviet Union. This shows just how much political influence these religious broadcasters have. Not only do they co-operate at high level with Western governments such as the US and Britain, but they have the necessary diplomatic muscle (and hard currency) to negotiate with Russia on leasing very high power SW installations located deep in the frozen wastes of Siberia for the purpose of broadcasting religious programmes to countries in Asia, Eastern Europe and to all parts of China.

The first SW facility that AWR deployed was at Novosibirsk, one of the many powerful SW transmitter stations the Soviets used during the cold war. The Novosibirsk complex was constructed in 1956, three years after Stalin's death. It comprises eighteen 100 kW SW transmitters and

Table 21.6 *AWR SW broadcasts from Russia*

Station	Power (kW)	Azimuth	Language	Target
Samara	250	188 deg	Arabic	Iraq, Gulf states
Samara	250	188 deg	Arabic	Iraq, Gulf states
Samara	250	188 deg	English	Iraq, Gulf states
Samara	250	284 deg	Swedish, Norwegian, Spanish	Scandinavia
Samara	250	284 deg	Polish	Poland
Novosibirsk	100	111 deg	Korean	Korea
Novosibirsk	100	111 deg	Mandarin	Northern China
Novosibirsk	200	178 deg	Mandarin	Western China
Novosibirsk	200	178 deg	Hindi	Northern India

numerous high gain curtain arrays extending over several hundreds of hectares. During the cold war years Soviet engineers devised SW stations with high gain arrays up to 8 wide × 8 high, capable of projecting high audibility SW signals over very long distances.

In 1993, taking advantage of further high power SW transmitter sites in Russia, AWR concluded arrangements to lease transmitters at sites near Samara, thus enabling it to target northern and western China and Korea. These arrangements continued until September 1994 by which time negotiations between AWR and the newly liberated Eastern European countries began to develop. These negotiations came none too soon because the Russian authorities, in the belief they had a captive market in high power SW complexes, had begun increasing leasing costs, and figures of up to $200 per transmitter-hour were being discussed. Late in 1993 AWR secured a lease on SW complexes in the Prague region at Rimavska Sobota, which had broadcast as Radio Prague during the cold war, and had powerful 250 kW transmitters operating into sophisticated 8 × 8 directive curtains.

The take-up of the lease by AWR on Rimavska Sobota did not signal the end of the relationship with Russia and AWR continues to have an agreement which enables it to use Russian sites.

With these former Soviet transmitters now targeting northern and western China, and the AWR-owned SW station in Guam targeting eastern China, there seems little doubt that the voice of AWR is generating a large audience in China.

FEBC: history and operational profile

The Far East Broadcasting Company (FEBC) was formed in California on 20th December 1945 by Robert H Bowman and John C Broger with the specific objective of broadcasting Christian radio programmes to China. Broger travelled to China in early 1946 but the Chinese government in Shanghai refused to grant a licence to build and operate radio stations within China. This reaction was perhaps not surprising due to the US government's support for the forces of Chiang Kai Shek.

Broger then went to Manila where in August 1946 the Far East Broadcasting Company (Philippines) Inc. was formed, while over in the US his partner Bowman was developing a base of financial support for the venture. The first radio broadcast from the Philippines took place on 4th June 1948 from a 1 kW MW transmitter in a paddy field north of Manila. It is very doubtful whether anyone in China heard that broadcast, and the real purpose of that first transmission was to satisfy the requirements of the licence. Later that same year the transmitting station was improved to enable a regular service to be sustained on both MW and SW.

By 1965, a time when the US government was opposing the communist forces by openly supporting the military build-up of Kuomintang (KMT) forces in Taiwan, FEBC began to build more transmitting stations not only in the Philippines but also on the American-administered Japanese island of Okinawa. By the 1970s FEBC had transmitter sites targeting China operating in Manila, Cebu, Bacolod, Zamboanga, Davao, South Cotabato, Legaspi, Zambales and Mindoro. On Okinawa two MW stations were opened which broadcast to China and the USSR. However, these transmitters ceased operation and were taken off-air when the island reverted to the Japanese.

Station KGEI, the 'Voice of Friendship', opened the Latin American service of FEBC in 1960. This station was located in Redwood City, on the salt flats of the San Francisco Bay. KGEI was originally built by the General Electric Company (GE) in time for the Golden Gate Exposition of 1939 and was later turned over to the US Office of War Information (OWI) during World War II when it was used to combat propaganda broadcasts from Tokyo. KGEI added a second 250 kW SW transmitter in 1973 to give FEBC better audibility in Latin America but also to enable broadcasts to Japan and the Soviet Union. The station was taken out of service in 1994.

In 1968 FEBC's associate company Far East Broadcasting Associates (FEBA) opened up a powerful SW service from the Seychelles with which to broadcast to the Middle East, Africa and parts of Asia.

FEBC's 'Open door to China' project has been described (see p. 179), initially broadcasting from Iba in the Philippines. To reach wider audiences in China, a 1 kW MW station was set up in the Northern Marianas, a US commonwealth territory. This was upgraded to 10 kW in 1984 and KFBS, the International Service of FEBC in Saipan, Northern Marianas went on air in 1984. Three more transmitters were added that year, all with 100 kW power, and a fourth 100 kW was added in 1987. The completion of this build-up of transmitter power marked the inauguration of 'Open door to China': Phase 3.

The transmitter network of FEBC/FEBA at the end of 1995 is given in Table 21.7. It has more than 30 transmitters in five different countries. Its staff has grown to 1000 and its annual operating budget is $17 million. In June 1996 FEBC had a weekly output of 631 programme hours and FEBA 142 hours – a total of 773 hours. It is interesting to note that this output is greater than that of some of the major international broadcasters such as Deutsche Welle (655) and Russia (720).

The FEBA SW station in the Seychelles

Approximately 1000 miles off the coast of East Africa, north of Madagascar, lie the Seychelles group of islands. Perhaps better known to Westerners for their holiday and tourist appeal, these islands have other uses. On the largest island of Mahe Far East Broadcasting Associates operates a SW station, as does the BBC World Service with its Indian Ocean Relay station with two 250 kW SW transmitters.

Table 21.7 *FEBC/FEBA transmitters and their target areas*

Transmitter designation	Location	Target area	Power	Frequency
FEBA (UK), 3 transmitters	Mahe, Seychelles	SE Africa, Middle East, India	100 kW	SW
KFBS, 4 transmitters	Marpi, Saipan	Russia, China	100 kW	SW
KSAI	Susupe, Saipan	Saipan local	10 kW	936 kHz
C-100	Iba, Philippines	Russia, China	100 kW	SW
Phil. OS (3)	Bocaue, Philippines	SE Asia, China	50 kW	SW
BSW-1	Bocaue, Philippines	SE Asia, India	100 kW	SW
GF-100	Iba, Philippines	SE Asia	100 kW	SW
HLAX	Cheju-do, S Korea	Russia, China	250 kW	1566 kHz
HLKX	Inchon, S Korea	N Korea, S China	100 kW	1188 kHz
HLAD	Taejeon, S Korea	Taejon local	3 kW	93.3 MHz
DZAS	Bocaue, Philippines	Philippines (all)	40 kW	702 kHz
DZFE	Karuhatan, Philippines	Manila area	10 kW	98.7 MHz
DWRF (2)	Iba, Philippines	Iba local	5 kW	1458 kHz
DXAS	Zamboaga, Philippines	Zamboanga local	5 kW	1116 kHz
DXFE	Davao, Philippines	Davao local	5 kW	1197 kHz
DXKI	Marbel, Philippines	Marbel local	5 kW	1062 kHz
DYFR	Cebu, Philippines	Cebu local	3 kW	98.7 MHz
DYVS	Bacolod, Philippines	Bacolod local	5 kW	1233 kHz
DWAS	Legaspi, Philippines	Lagaspi local	5 kW	1125 kHz

A British possession for nearly 200 years, administered from Mauritius, the Seychelles became a separate British crown colony in 1897. Now part of the British commonwealth, political stability is ensured, and in every other respect the islands are an ideal choice for a SW station site. The climate is temperate, ground conductivity is very good, and the surrounding Indian Ocean makes for ideal SW broadcasting. Moreover, the presence of coral reefs makes it practical to have the SW curtain antenna arrays erected on a water table with no obstructions in the foreground. The strategic location is ideal for projecting signals with high audibility to the whole of East Africa, from South Africa to Ethiopia and Somalia, as well as to the Indian sub-continent, Pakistan and parts of the Middle East. All these regions FEBA Radio reaches from its SW transmitter site at Mahe.

FEBE operates three 100 kW SW transmitters; a Gates (Harris) SW 100 from 1973, a Continental Electronics 418C-2 from 1979 and a Harris SW 100B from 1988. Through a switch matrix any transmitter can be connected to any one of seven antennas. These are all highly directive curtain arrays with the following characteristics:

- Two antennas centred on azimuth bearing 280 deg:
  HR 2/23.6/.7   15–17 MHz
  HR 2/2/.6/.7   9–11 MHz

- Three antennas centred on azimuth bearing 040 and 220 deg, slewable ±12 deg.

  HRR 2/2.6/.7   6– 7 MHz
  HRRS 4/2/.6/.7   9–11 MHz
  HRRS 4/2/.6/.7   15–17 MHz

- Two antennas centred on azimuth bearing 340 deg., slewable ±12 deg.

  HRS 4/2/.6/.7   9–11 MHz
  HRS 4/2/.6/.7   15–17 MHz

  All antennas are of Brown Boveri Company design, FEBA construction.

An application had first been made by FEBA for the site in 1969. By 1970 the building of staff houses and accommodation for studios and a transmitter was complete. The first SW transmitter was a 25 kW ex-navy transmitter but it was a start and in May 1970 it went on air for the first time carrying 'the voice of God' (in FEBA's words) to India. That same year FEBA embarked on the construction of its reef antenna project, which cost some $0.5 million but was capital well spent. Today the same antenna system, with its SW curtains erected on a coral reef in the sea, would cost nearly twenty times that figure. This antenna system, with seven SW directive curtain arrays, is the key element of this SW transmitting station.

1979 saw the installation of the second 100 kW SW transmitter,and nine years later a third 200 kW SW transmitter was added because the ageing original 25 kW transmitter has been retired from service. Douglas Malton, the chairman of FEBA Radio, is enthusiastic about the effectiveness of SW radio and highlights one country alone to which FEBA radio broadcasts –

Figure 21.3 *Reef antenna system, FEBA, Seychelles*

India, with its population of over 900 million. To put Christian missionaries on the ground in India would cost an immense figure, taking into account the numbers that would be needed, but SW broadcasting can reduce drastically the number needed and its effectiveness is proven. Today, FEBA Radio broadcasts around the clock, taking the gospel to listeners in more than 30 countries in 45 different languages; an achievement that would be a credit even to a government-funded international broadcaster.

For its international broadcasting operations FEBA, as with national religious broadcasters in the US, claims to be funded by voluntary donations from Christian believers, which it solicits in good measure. It claims not to receive money for commercial advertising, or any government agency, but its programme suppliers contribute to the cost of the broadcasting of these programmes, a fact that makes FEBA a surrogate broadcaster.

Transmitter manufacturers, in particular Continental Electronics and Harris, also have reason to be grateful to such religious broadcasters. My analysis of SW transmitter sales puts these two companies at the top of the sales league. Both companies have the distinction of supplying SW transmitters to TWR as well as FEBC and its British associate FEBA Radio, and WYFR (World Family Radio). Continental has supplied the Christian Science Monitor with SW transmitters, whilst Harris has sold SW transmitters to stations KWHR, WWCR (Worldwide Christian Radio), WHRI (World Harvest Radio) and KTBN Salt Lake City, all in the 100 kW transmitter power bracket.

Figure 21.4 *Transmission line feeders and reef antennas, FEBA, Seychelles (antennas up to 100 m tall)*

Chapter 22
The broadcast transmitter industry

The 1990s has not only seen some momentous changes in world politics, it has seen a dramatic change take place in the transmitter manufacturing companies in Europe and America.

From the 1960s, with their manufacturing facilities fully recovered from the effects of World War II, the giant electrical companies of Europe were in the ascendancy. Europe was the birthplace of radio broadcasting so it was appropriate that these electrical companies, which included Brown Boveri of Switzerland, GEC Marconi of England, AEG Telefunken and Siemens of Germany, and Thomson of France, should come to dominate the high power sector for SW.

All these companies had contributed in some measure to developments in high power, and the key to their extraordinary success was competition, the driving force that generated technical excellence. Each company managed to dominate the markets at different times.

In the late 1970s Telefunken led the rest in its pursuit of better overall electrical efficiency; it developed the world's first 500 kW SW transmitter with Pulse Duration Modulation (PDM), thus raising transmitter efficiency from 55 per cent to 65 per cent. In 1984 Brown Boveri introduced its revolutionary Pulse Step Modulation (PSM), offering an increase in overall electrical efficiency to 75 per cent. From 1985 onwards, Brown Boveri has dominated the world market, selling to international broadcasters in the USA, Europe, the Middle East and Asia (see Appendix 2).

Brown Boveri was also one of the world leaders in the construction of very high power MW and LW transmitters. Such was its domination in all sectors of radio broadcasting, Brown Boveri could afford to design and construct its own range of high efficiency power grid tubes rather than rely upon other manufacturers. To the people who worked at its transmitter factory in Turgi, Switzerland it must have seemed that their good fortune would go on and on. Then in February 1992 came the surprise announcement from ABB headquarters in Zurich that it had decided to dispose of its high power broadcast transmitter businesses in order to concentrate upon

191

its core businesses in energy generation and traction. The named buyer was Thomson-CSF, its main rival in the transmitter business.

To this author and many others in the broadcast transmitter industry it was almost unbelievable that the parent company ABB, in neutral Switzerland, should even contemplate selling its broadcaster transmitter business, which had held an unequalled reputation for building high quality transmitters of outstanding designs and long life. Many in the broadcast transmitter industry believed that it would have made more sense if ABB had bought the transmitter business of Thomson-CSF because ABB's share of world markets was then 37 per cent compared to the 22 per cent of Thomson-CSF over the same period (1985–1991).

And so, by the end of 1993 the number of transmitter manufacturers in Europe had shrunk to three major companies, GEC Marconi, Telefunken and Thomcast (formed from ABB and Thomson-CSF). However, RIZ in Zagreb had by now been privatised and set its sights on being another major player.

From the 1960s, the US, by comparison, had seen the decline into obscurity of famous names such as RCA, Hughes, Collins and General Electric so far as transmitter technology was concerned, leaving just the two companies Harris and Continental. In the late 1980s, however, the Broadcast Transmitter Division of Harris Corporation had lain a groundplan to capture world markets for high power MW. When Harris in 1989 ran a small advertisement requesting enquiries for its 500 kW MW transmitters, using solid state DX technology, few in Europe took it seriously. One European manufacturer expressed an opinion to this author that Harris simply did not have the experience but soon afterwards (March 1992) it demonstrated the world's first all solid state high power MW transmitter at the plant in Quincy, Illinois. From then on, the company built and sold solid state DX transmitters with carrier powers up to 400 kW, then 600 kW and 1000 kW, and finally rounded off in 1995 by winning the tender to supply the oil state of Qatar with a two megawatt system. Meanwhile the Continental Electronics Corporation had invested much effort into producing an entirely new transmitter range for LW, MW and SW. Both these companies were aiming to reduce the domination which European companies had exercised for so long.

Strategic alliances: a key to global markets

The world markets for the supply of high power broadcast transmitters and the execution of very large turnkey projects are becoming increasingly competitive. This has been brought about by two main factors; first, there is little expansion in the market – much of the business is that of replacement – and, second, the great speed with which companies like ABB completed major turnkey projects, coupled with faster production times for high power transmitters.

In the same way that the major players in the space industry and in other sectors are forming strategic alliances with former rivals, so the broadcast industries of the US and Europe are now undergoing some rationalisation. There are sound economic reasons for the forging of alliances, whether they are takeovers, mergers or simply collaborative marketing. Alliances offer many cost-reduction advantages, in areas ranging from research and development, production, marketing, right through to after-sales service. Also, where an element of cross-transfer of technology is involved, this can actually result in a better product at a lower cost.

In 1995 Daimler Benz Aerospace took the decision to concentrate in future on its core business, which resulted in AEG Telefunken being offered for sale. Telefunken is one of the most famous names in the history of broadcasting, and it seemed to some unrealistic that the board of Daimler Benz should even consider disposing of it. However, it did and in December 1995 it was acquired by Tech-Sym Corporation of Houston, Texas for an undisclosed sum. Thus Telefunken became a German subsidiary of Continental Electronics Corporation, which itself was a wholly owned subsidiary of Tech-Sym Corporation. It is open to speculation why Thomcast did not step in to bid for its old rival Telefunken Sendertechnik.

Thus the Continental Electronics Group, comprising Continental Electronics and its subsidiaries Telefunken Sendertechnik GmbH in Berlin and Continental Lensa SA in Santiago, Chile, now effectively constitutes one of the largest groups in the broadcast and communications industry, rivalling those of Thomcast in France and the Harris Corporation in the US.

Major players in the broadcast industry in North America

The broadcast business is a global industry as anyone who has attended a broadcasting convention will know. This section confines itself to those companies which compete with Continental and its subsidiaries. Continental Electronics and Telefunken Sendertechnik are multi-disciplinary companies whose core business lies in SW and high power AM but whose other business includes FM, digital and analogue TV and DAB transmitters. In the US, Harris is its only major competitor but does not manufacture high power SW transmitters.

However, for high power AM, MW and LW broadcasting Harris is a formidable competitor and by market sales it is world leader. Nautel in Canada is a much respected company in the AM market, and pioneered solid state AM before Harris, but has not moved beyond the 300 kW carrier power level. Broadcast Electronics in Quincy, Illinois is another excellent company but it competes in a fairly narrow sector of broadcast technologies.

Chapter 23
Company profiles

GEC Marconi

The Marconi Company, as it was called a few decades ago, was a company born with the advantage of holding Guglielmo Marconi's patent on wireless. World War II saw the factory at Chelmsford in the UK busy turning out equipment for communications and broadcasting. At the end of the war Marconi's factories were intact, as was its labour force, and it had access to world markets denied to other European companies like Philips of the Netherlands, Thomson of France and particularly Siemens and Telefunken of Germany.

Marconi entered the post-war years with all the confidence of having been on the winning side. This confidence was reflected in the sales of its radio broadcast transmitters. From 1950 to 1955 its sales exceeded those of all its competitors with the exception of the Brown Boveri Company. It is significant that Brown Boveri in neutral Switzerland did not suffer any ill-effects from World War II, and its factory remained intact. In the period from 1960 to 1970 Marconi was at full steam with its sales of broadcast transmitters exceeding those of Brown Boveri and its other competitors, AEG Telefunken and Thomson-CSF. In this period Marconi delivered transmitters to many countries around the world, including many to Britain's commonwealth countries. It sold sixteen of its BD 253, a 100 kW SW transmitter, while the BD 272, a 250 kW SW transmitter, went on to clock up a remarkable thirty sales by 1967.

By the late 1970s, however, Marconi had slumped from the top of the league table to the lower end, in terms of measured sales, a position from which it might never have recovered had it not been awarded the VOA modernisation programme tender. As one senior engineer at Chelmsford put it, 'Whether we lost money on that VOA Modernisation Programme is now history. The important thing was that it got us back into the broadcast transmitter business.'

Before winning the VOA contract the company had sold four of its B 6127, its first full 500 kW transmitter with pulse modulation. The B 6128, the successor to the B 6127, sold two beyond sales to VOA, whilst the B 6132, Marconi's first 500 kW SW transmitter with a solid state modulator, sold just one unit.

Ascribing reasons for company success, or the lack of it, is not easy because sales orders can be influenced by any one of a number of factors, and not just production cost. It is known that production costs were higher in Germany and Switzerland, yet sales from the manufacturers in those countries outperformed those of Marconi. A more likely factor to account for its poor performance is that Marconi appears to be driven by evolutionary strides, rather than by highly innovative engineering. Many high performing companies are those that have made technological breakthroughs. For example, Marconi did not build a 500 kW SW transmitter until 1984, more than a decade after its competitors. AEG developed PDM, ABB developed PSM, and Thomson excelled in producing 500 kW transmitters using one tube in the final stage. Practising safe evolutionary design is a lot cheaper than practising innovative engineering, but it does not make for world-beating sales. On 11th June 1997 GEC-Marconi Communications announced that it was withdrawing from the high power broadcasting and antenna business, a century after its entry. Britain thereby ceased to be a manufacturer of high power transmitters and exited a potentially valuable export market, as has been the case in many manufacturing sectors.

RIZ: Radio Industries Zagreb

The foundations of Radio Industries Zagreb (RIZ) were laid in 1948, as a result of the need for a radio electronics industry identified by the Socialist Federal Republic of Yugoslavia. RIZ was founded by a team of experts from Zagreb Radio with the intention of producing studio equipment to meet their own needs.

RIZ was a partner of a state authority called the Composite Organisation of Associated Labour and was founded by a decree of the government of the People's Republic of Croatia, on 24th December 1948. Initially there were about 50 workers, yet such was the startling pace of growth in this state-owned enterprise that 20 years later the company had 2821 employees, and by 1979 that figure had risen to more than 4000.

Although RIZ came from small origins, the rapid speed of development in electronics from the 1950s onwards was such that RIZ came to be a major player in the economy of Croatia and indeed Yugoslavia. RIZ subsequently grew into five works Organisations: RIZ Electronics, Television and Acoustics; RIZ Professional Electronics; RIZ Transmitter Factory; RIZ Industrial Electronics and RIZ Production, Research and Development of Parts and Components.

Figure 23.1 *300 kW mobile MW station under construction, RIZ*

In essence, RIZ developed and manufactured anything and everything to do with radio broadcasting, industrial electronics, telecommunications, factory automation, railway signalling systems, domestic radio and TV sets and supplied military equipment to the Yugoslav People's Army. Of the different RIZ companies, the transmitter factory came to play a key role in various market sectors. The first of these in equipping the Republic with

Figure 23.2 *100 kW mobile MW radio station broadcasting for HRT*

Figure 23.3 *100 kW SW PSM transmitter, RIZ works*

radio transmitters for AM broadcasting and later, in collaboration with Siemens, TV broadcasting. Another was in telecommunications and a third was in the transmitter export market. RIZ won valuable contracts to supply high power broadcast transmitters to countries in Africa and the Arab world.

Yugoslavia was unique in the political world as the only neutral communist state. Under Tito's leadership it practised separatism and centralism at the same time; Tito recognised the ethnic and political divisions between the various Yugoslav republics, yet at the same time managed to hold the Federation together right up to his death in 1980. Yugoslavia was unique in other ways too; although a socialist state it was not closed to the outside world, unlike the USSR, and it succeeded in riding a fine line between socialism and capitalism. Some proof of this achievement is that the RIZ company forged partnerships and trading agreements with many major manufacturers located in the capitalist world, such as General Electric, Rockwell, Fairchild, LM Ericsson, Philips, TCI, Telefunken, Siemens and others. This cross-fertilisation of business relations was beneficial to both sides and played a large part in the growth and evolution of the RIZ companies.

The RIZ transmitter factory, for example, created a very successful business co-operation with Siemens of Germany. This enabled a cross-transfer of ideas and transmitter technology such that by the end of the 1960s RIZ was a strong player in the business of bidding for major high power transmitter projects around the world. Transmitter production began initially with output powers of 2–20 kW in the early period up to 1950, but by 1968 the company was able to deliver 600 kW MW transmitters to Bavarian Radio, Munich. Also in 1968, the company built and delivered its first 1000 kW

MW transmitter to India, and ten years later another important milestone was set when the company won its first major turnkey project — a two megawatt MW broadcasting station in Libya. By any standards this growth was a tremendous achievement. Only two other manufacturers in Europe had built radio transmitters of such power and only one other company (Brown Boveri) could equal the record set by RIZ in its sales to the Arab world. Within a period of less than 30 years the RIZ Transmitter Company had progressed from a fledgling enterprise to become a major European transmitter manufacturer, second only to Thomcast.

Today RIZ has three main activities: its stationary programme, the design and manufacture of high power broadcast transmitters for fixed installations; its mobile programme, the design and construction of broadcast transmitters for mobile/transportable applications; and the planning and execution of major projects.

Mobile broadcasting centres

In the 1970s RIZ began building mobile radio and communications systems. Previously the main users of mobile or transportable communication systems had been military forces and NATO in particular. RIZ offered transportable systems with MW and SW transmitters to the civil market. Over the years RIZ has acquired a lot of experience in this highly specialised field, which calls for knowledge of how to design out the effects of shock and vibration, which can often render electronic equipment and tubes useless.

Television and radio have evolved to become powerful media and it is not surprising that in times of war or dispute, radio and TV stations are early targets. In the 1990s there have been three conflicts in which the first military targets were radio and TV stations; the Gulf War, the war between Croatia and Serbia and the recent NATO action against Serbia.

Apart from the obvious advantages of mobility there are other significant benefits from having a transportable radio station; it can be assembled, wired, factory-tested and even air-tested before despatch to the customer.

The market for transportable systems falls into three categories:

- Medium power AM systems up top 50 kW, single and dualled;
- TV transmitting stations up to 40 kW peak sync power for broadcasting in television bands III, IV, and V;
- High power AM stations up to 300 kW carrier output power.

Such stations are usually built as self-contained systems with one or more standard size ISO containers, complete with water for transmitter cooling, input racks, switching systems, dummy loads, masts, antennas and feeders. Programme feeds might be by microwave, UHF link or by down-link satellite earth station using a modest-sized dish.

A number of companies have built systems in the first two categories, but RIZ outdistanced other manufacturers by being the first company in the world to build high power AM transmitter systems with carrier powers up to 300 kW. In 1998 the company secured its first order for a 300 kW MW transportable broadcasting station. This came from the US government and its role remains secret. This 300 kW transmitter system, designated the OR 300 SD-1M and built by RIZ in Zagreb, uses Harris's DX technology and power modules. The extremely high electrical efficiency of the OR 300, 80 per cent or better, makes it possible to extract heat from the containerised system, allowing the transmitter to perform in harsh environments: ambient temperatures up to 50° C, and at altitudes up to 4000 m, with a derating of 6 degrees per 1000 m of altitude. Clearly, this 300 kW mobile radio station represents a significant milestone in the evolution of mobile broadcast transmitters, and its ability to alter frequency in a matter of minutes makes it suitable for jamming operations.

Like most other transmitter manufacturers, RIZ has collaborative ventures with other companies. It uses Siemens tubes wherever possible, from a conviction that Siemens tubes possess long life and give outstanding performance. The RIZ deal with Harris gives Harris another outlet for its DX technology, selling its basic modules to RIZ, and a foothold into wider markets. In return, RIZ gains from more orders for its transmitters and mobile systems. Harris has even turned over some of its SW patents to RIZ. Such a collaborative venture offers excellent prospects to both companies.

The war between Croatia and Serbia in 1991, and continued tensions in the Balkan region, have an effect on RIZ's ability to win large contracts. Broadcasting authorities want assurance that the contractor has a stable future.

Telefunken Sendertechnik

By any standards Telefunken is a unique company. Its products span the entire spectrum of communications, radio broadcasting and television and newer technologies like digital broadcasting. Its expertise extends to sophisticated antennas whose designs call for much knowledge in electromagnetic science. Only Thomcast of France can match the capabilities of this Berlin-based company.

But Telefunken's history is just as impressive as its products, and its plant in Berlin is every bit as famous as its English counterpart Marconi. Both companies were at the birth of spark communications and high power LW telegraphy, before the birth of sound broadcasting. Telefunken, formed in 1903, was a result of a consortium of AEG and Siemens and made history from the start; in 1906 it established a record in wireless communications when it transmitted signals over a path length of 3600 km from its site at Nauen.

When the short waves were discovered Nauen was equipped with SW antennas and transmitters by Telefunken. During the cold war, from 1949 Nauen became a town in the German Democratic Republic and the SW transmitting station played a key role in the propaganda war between East and West. The happy ending to this piece of history is that the site is now used by Germany's international broadcaster Deutsche Welle.

In December 1995 Telefunken was acquired from Daimler Benz Aerospace AG by Tech Sym Corporation, the owners of Continental Electronics Corporation of Dallas, US, for $9 million. It remains a Germany company, as a subsidiary of Continental. Why the company was not bought by Thomson is something of a mystery, but the most likely explanation is that the company came up for sale at the time when Thomcast had not completed its rationalisation of the ABB transmitter business it had acquired two years earlier.

The sale of Telefunken was a big loss for Europe, as all the indications are that it will continue to play a key role in all sectors of radio and TV broadcasting, including emerging technologies like Digital Audio Broadcasting (DAB). For Continental the acquisition makes sound logic. Although both companies manufacture high power SW transmitters with powers that overlap, thereafter the two companies' products are complementary. Telefunken alone has an antennas and structures division and its range of television fills the gap in Continental's product ranges.

Figure 23.4 *Telefunken S4105 500 kW SW transmitter installed in the blockhouse of the new Nauen SW centre*

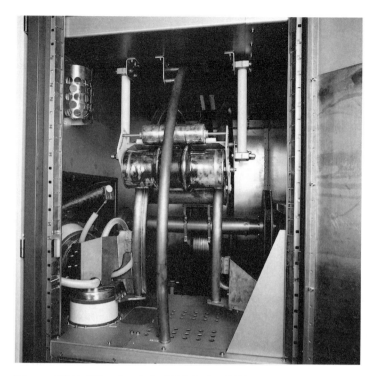

Figure 23.5 *RF tuning deck of Telefunken S4105 transmitter*

Telefunken has been a truly innovative company with 'firsts' in antenna design and SW transmitter technology with DAM and PDM. Telefunken's commitment to investing in long-term research and development is deep rooted and an important part of the company character. It has enabled the company to stay at the forefront of emerging technologies, and in many instances to be the leader, such as in the development of DAB technology.

Within the AM sector of radio broadcasting, in the first half of the 1990s Telefunken has sold fifteen 500 kW SW transmitters, 22 high power MW transmitters and more than 70 high power antennas, including two 500 kW rotatables. For a broadcast medium whose demise has been predicted since FM broadcasting was introduced in the early 1970s, such sales figures are quite remarkable. The figures are even more impressive if account is taken of what high power AM transmitters used to cost – a 500 kW SW transmitter would cost $2 million. Today's improved designs have lowered these costs, but they are still higher than for a high powered UHF television transmitter.[1]

---

[1] On 9th July 1998 Tech Sym Corporation announced that it intended to sell several businesses to improve overall performance and sharpen its corporate focus, including the broadcast communication business unit which comprises Continental Electronics and its subsidiaries Telefunken Sendertechnik GmbH in Berlin and Continental–Lensa SA in Chile. At the time this book went to press no buyers had been reported.

Thomcast

In 1993 the French-based Thomson-CSF took over the broadcast manufacturing business of former rival Asea Brown Boveri, which had decided to concentrate on its core business of energy engineering. Thomson-CSF merged the operations into Thomcast, a wholly owned subsidiary, which is now the largest such manufacturer in the world.

ABB and Thomson were both world leaders in transmitter technology, producing powerful television and radio stations across (literally) the whole spectrum, from UHF to LW. They were well matched in expertise, experience and resources. The merger made Thomcast a European giant, with a vital size advantage over its competitors.

Radio broadcasting is a highly developed science where continuing investment in research and development is a vital ingredient. Both Thomson and ABB had always invested heavily, but independently; now the combination of their two activities could lead to a cross-fertilisation of ideas.

Broadcasting technology has matured to the point where each significant advance becomes more costly to bring about – the classic case of diminishing returns. In SW broadcasting, for example, some stations now produce 1000 kW of power from two co-phased antennas with such high power the classical form of a SW transmitting station is no longer appropriate, and a radical rethinking becomes necessary to address issues such as the ecological effects of highly concentrated radio waves.

Thomson-CSF was the first to address this problem, with the ALLISS design. The first ALLISS SW station, which went on air for Radio France International in 1995, marked the first active collaboration of ABB and Thomson-CSF engineers.

Figure 23.6 *Two 1000 kW LW transmitters coupled at Europe No. 1 Broadcasting Centre, by Thomson-CSF*

Thomcast is based at Conflans Sainte Honorine, near Paris, which houses the administrative headquarters and the manufacturing site of Thomcast France. The new company will make equipment on this site ranging from VHF/UHF TV transmitters, 100–1000 kW MW transmitters to 2000 kW transmitters at very low frequency and LW. For SW broadcasting the range is from 100 to 500 kW.

Associated with these products and disciplines is logistic support and turnkey capability. Conflans also has a strong capability in designing antenna systems across the entire spectrum of TV and radio broadcasting.

Thomcast AG Switzerland, formerly ABB Infocom, has long been associated with powerful AM stations. Since introducing its series of PSM transmitters in 1984 the company has sold more high power transmitters than any other. With an enviable reputation for reliability and long life, ABB was the market leader in the Middle East and Gulf region; according to my analysis, it had captured 40 per cent of this market since 1986.

Thomcast GmbH Germany, previously ABB Leitungsbau Antenna GmbH, in Mannheim, is an international force in the highly specialised field of antenna design, fabrication and erection. For SW broadcasters who want flexibility to address any region of the world, it has developed a sophisticated design of rotatable curtain antenna that is already in service with many broadcasters. The company will play a key role in providing antennas and structures for the whole Thomcast group.

ABB had a reputation for customer service, and Thomcast needs to ensure the loyalty of its customers by upholding ABB's service commitments, such as provision of spares regardless of the age of the transmitter, and services to any installation.

At the time of the merger there was some overcapacity in the market due to two factors: political changes in certain parts of the world, affecting the high-power SW market, and the world recession, affecting all sectors of business. However, Thomcast's long-term view is that radio broadcasting in all sectors using terrestrial transmitters will survive well into the twenty-first century.

Another factor that led to overcapacity could have been the efficiency and speed with which the companies execute major turnkey contracts. It used to take 3–4 years, excluding the planning stages, to build a high-power SW station; but ABB Infocom completed a recent project in Jordan in 18 months. Now the same plant in Turgi, Switzerland, has devised a way to reduce total project time to less than six months. This is good news for customers, but puts more pressure on the other suppliers. Thomcast now has the resources and expertise to more than hold its dominant position in the global market.

Continental Electronics Corporation

Continental is the youngest of the major manufacturers of high power broadcast transmitters. It was established in Dallas, Texas in 1946 by James Weldon with the goal of becoming pre-eminent in its chosen field; the generation of radio frequency power on a scale which at that time had not been achieved by any other company.

Although much of the success that Continental has enjoyed has come about because it was created in the right place (the US), at the right time, and in a discipline which had a future, no one can deny that Continental's rise to rival the major manufacturers in Europe was impressive. By 1951, less than five years after being established, the company had built eight 500 kW SW transmitters for Voice of America, admittedly by combining pairs of 250 kW transmitters but still a great achievement. Three years later the company built three 1000 kW MW transmitters using the same technique of paralleling two 500 kW units, and between 1973 and 1981 built eight 2000 kW MW transmitters, the highest powered transmitters in the world at that time.

All these orders came directly or indirectly from the US government, for the cold war or America's other involvements in Europe and Asia. Continental's engineers have excelled at innovative engineering on scales not previously attempted by other companies, and the company's commitment to excellence extends across the radio frequency spectrum, from ELF (extremely low frequency) to UHF, S Band and beyond, with transmitter powers from kilowatts to megawatts.

No other US transmitter manufacturer served its country so well during the forty year cold war. It was by far the biggest supplier of high power transmitters for US SW and MW broadcasting to Soviet controlled countries. Nor has any other manufacturer equalled the length of service which the continental transmitters have clocked up. Its 500 kW SW transmitter supplied to VOA's Greenville station in 1951 is still in service. Only in recent years have the forty year old 1000 kW MW transmitters in the VOA stations in the Philippines and in Thailand been replaced with new transmitters – not, it should be stated, because of unreliable service, but simply because these old transmitters were expensive to run due to their lower electrical efficiency.

Continental has been the major supplier of high power SW transmitters to Radio Free Europe/Radio Liberty. Fifteen of its 250 kW transmitters are in service in one station alone, the Gloria SW facility in Portugal. Such power, combined with very high gain in curtain arrays, produces a devastating amount of radio frequency energy. However, it has been in the 100 kW power level where the company has excelled; an analysis of SW transmitters sales between 1990 and 1996 shows that Continental captured 52 per cent of total world market with forty transmitters delivered.

The company has not been without its share of upheavals or disappoint-ments. In May 1985 Continental was purchased by Varian of California, the manufacturer of tubes (for which Continental was one of its biggest customers). Five years later Varian sold off its transmitter manufacturing interests in Dallas and Cambridge, UK, and Continental then became a wholly owned subsidiary of Tech Sym Corporation. In 1989 the company failed to win the order from Voice of America to re-equip its SW relay network, losing it to GEC Marconi. This was a major disappointment to Continental and Asea Brown Boveri, both having been tipped as having the best chances of success.

After this set-back, Continental went ahead with an intensive develop-ment programme on a new series of AM transmitters for SW, MW and LW broadcasting. For SW, these were types 418E 100 kW, 419G 300 kW, and 420C, the 500 kW model. A rationalised design approach ensured that many units were common to the entire transmitter range, with the principle differences being in types of tube, rating of vacuum capacitors and in switchgear.

Since 1993 Continental has been successful in selling its high power broadcast transmitters to a number of religious and international broad-casters. The US International Bureau of Broadcasting (IBB) standardised in 1998 on the procurement of Continental's SSM modulator for retrofitting to a number of its overseas SW stations.

In December 1995 the fortunes of Continental took another giant leap when its parent company Tech Sym Corporation announced the purchase of Telefunken Sendertechnik. Although Telefunken will stay as a separate company, there will undoubtedly be major benefits from the close alliance of two major players.

One of the reasons for success of Continental is that it recognised that to be successful it had to concentrate upon the transmitter markets rather than attempt to compete in a wide range of disciplines. Thus it was convenient to work with and use antennas from companies such as TCI. However, the acquisition of Telefunken gives Continental a capability in antenna design and manufacture which has few equals plus Telefunken's ability and experi-ence in the planning and execution of turnkey projects.[1]

The Gates Radio Company (later the Harris Corporation)

In *History of international broadcasting* [2] the history of this pioneer broadcast equipment manufacturer, formed at the birth of US radio broadcasting, was left out, an omission rectified here.

[1] On 9th July 1998 Tech Sym Corporation announced that it intended to sell several busi-nesses to improve overall performance and sharpen its corporate focus, including the broadcast communication business unit which comprises Continental Electronics and its subsidiaries Telefunken Sendertechnik GmbH in Berlin and Continental–Lensa SA in Chile. At the time this book went to press no buyers had been reported.

The background of the birth of the Gates Radio Company can be traced back to the early years of experimental broadcasting in the US. Following the end of World War I there began a flurry of activity in wireless broadcasting, as it was then called. Hundreds of radio amateurs and other experimenters began to construct wireless stations. They were being built in garages, coal sheds, attics and even in chicken coops. The pattern was much the same; these experimenters acted as station engineers, announcers and sometimes as soloists accompanied by music from phonographs. All these wireless stations were unlicensed, a situation that changed when radio station KDKA came up on air with call letters issued on 27th October 1920. KDKA made history that year when it ran a broadcast on the Harding–Cox presidential election returns and was able to announce at midnight that Harding had won. It is doubtful whether more than a few hundred people heard the broadcast because mass production of wireless receivers had not then begun, but the radio station in Pittsburgh created a tidal wave of interest and excitement that spread far beyond that city to other mid-American towns and cities.

As is often the case with new technology, it is youth which first embraces it. One of those whose interest was fired was young Parker S Gates, teenage son of Henry C Gates, who lived in the mid-American small town of Quincy, Illinois. Henry Gates was assigned to work in Pittsburgh in 1921 where he accidentally came into contact with KDKA. He became interested as he foresaw a potential new industry and an ideal fit for his son Parker, whose hobby was radio.

In 1922 Henry Gates relocated his family back to Quincy and formed the Gates Radio Company with himself and his wife as joint stockholders and with their son as the designer of wireless receivers. It was not long before Parker Gates began to demonstrate a flair as a product innovator and by the late 1920s the company introduced a range of audio equipment for radio stations.

Henry Gates died in 1934 and Parker Gates became general manager. By 1939 the company produced its first broadcast transmitter, a 250 W AM unit for a radio station in Iron Mountain, Michigan. When America entered World War II, it brought about huge growth in government contracts and Gates Radio Company became an important RCA subcontractor. It was in this period that Gates made its most dramatic expansion producing radio equipment for the Armed Services from the factory in Quincy. This included the RCA Model ET-4750 7.5 kW SW broadcast transmitter, which was also supplied under a lease-lend programme to British Armed Forces, the BBC and to the Soviet Union.

Such was the rate of expansion of the company that by 1945 it had moved to its fourth location, although still in Quincy. With the end of the war Parker Gates quickly saw the coming demand for broadcast equipment, recognised the value of the RCA subcontracting experience and marshalled all company resources to develop a full line of radio broadcasting equipment to meet the needs of the new AM radio stations which were being licensed in towns and cities all over the US.

At that time its main competitors were RCA, Western Electric, GE, Collins, Raytheon and Westinghouse. During the next decade Western Electric, Raytheon and Westinghouse would drop broadcast products. (Later, in 1972, GE Broadcast Division would be acquired by Harris and RCA would finally depart from a sector of business in which they had always been the leader.)

On 1st November 1957 the Harris Intertype Corporation acquired Gates Radio Company. It was a shrewd move by a manufacturer whose previous experience was entirely in printing machinery. Not only did Gates possess an enviable track record, it gave Harris Intertype Corporation a technical resource for electronic typesetting and a foothold in a new industry with worldwide markets. As the *Wall Street Journal* reported, 'The move was particularly important since the application of electronics is becoming more and more important in the printing field'. At the time of the acquisition Gates employed 400 people with annual sales running at approximately $6 million (6th November 1957).

By the mid-1960s Gates had established itself as a major manufacturer of studio and MW transmitting equipment and become an important supplier of SW broadcast transmitters, not only to the US government, for the equipping of the VOA station at Greenville, but also to a number of countries in SE Asia. In 1962 Gates developed a new concept of a helicopter-transportable high power broadcasting station. Two 50 kW transmitters, one MW and one SW, were installed in standard US Army S141 shelters for rapid deployment by helicopters for the Army Psychological Warfare unit. The designer of the Gates transmitters was Les Petery who had joined Gates in 1946, having previously worked for the Crosley Radio Corporation at its SW station in Bethany, Ohio. A prominent transmitter designer, he had gained his experience in an era when it had become common practice for broadcasting companies to design and build their own transmitters and transmitter stations.

Gates' ability as a major manufacturer of radio transmitters continued to expand apace with a range that extended from FM, solid state, and high power AM to SW. In 1973 the company installed its first 50 kW PDM MW broadcast transmitter at KDKA Pittsburgh, having previously developed the world's first PDM transmitter. This was the VP100, a 100 kW model, the first of which was installed at Chang Mai, Thailand and resulted in a US patent being issued to designer Hilmer Swanson.

Harris ran Gates as a separate subsidiary for almost twenty years. During that time Parker S Gates remained active as President until late 1967, succeeded then by Lawrence Cervon as Vice President–General Manager. Under the leadership of these two veterans of the US broadcast manufacturing industry, Gates Radio Company continued to win valuable contracts from the US Information Agency (USIA) for 50 kW SW

Figure 23.7 *DX transmitter under test at Harris Broadcast, Quincy, IL, US with (l. to r.) TE Yingst, H Swanson and the author*

transmitters to be deployed by VOA in Rhodes, Greece, Liberia, the Philippines and Greenville, North Carolina, in the ceaseless battle for peoples' minds during the cold war. Cervon was an outspoken advocate of the theory 'that the best way to combat Soviet jamming was to send a stronger signal'. At that period of the cold war (the 1960s), 100 kW carrier power represented modern high power transmitter design, and the SW transmitters from Gates were fully state-of-the-art.

Figure 23.8 *Lawrence J Cervon with the author, at Quincy, IL, US*

Not until 1974 did Harris Intertype Corporation think it fit to drop the name Gates, simply referring to the organisation as the Harris Broadcast Division. Even then, the letter sent to customers notifying them of this change seemed almost to apologise, as if Harris was unsure of the reaction – 'Although the name has changed, I want to assure you that the company has not. We plan to continue to serve you as in the past, with the finest broadcast lines available.'

By this time the two men who had headed and guided Gates Radio company were no longer with the company. By then, Parker Gates had retired and Cervon had left Harris,[1] subsequently to become President and a major shareholder of Broadcast Electronics Inc., which he had moved to Quincy from Silver Spring, Maryland, where it was founded in 1969.

Parker Gates left a legacy to the broadcasting and communications industry which lives on and has emerged to be one of the world's most prominent broadcast transmitter enterprises. More than that, Parker Gates laid the foundations for the small mid-American city of Quincy, located on the banks of the Mississippi River, to become a highly diversified world centre of excellence in electronics manufacturing. Archival records show that by 1969 over 2000 radio stations within the US alone had bought Gates broadcast transmitters. Today, Harris products for radio broadcasting are to be found in more than 100 countries.

Harris Broadcast Division

Harris Broadcast Division is part of the Harris corporation, a $3 billion enterprise which is focused on four major businesses which range over the entire spectrum of electronics, communications and broadcasting. The traditions set by the Gates Radio Company live on, and today the Harris Broadcast Division has an unequalled record as an innovative company and for recognising new markets before most other companies. Its number of 'firsts' include the world's first all solid state FM exciter (1967), the first superpower 220 kW UHF-TV transmitter (1970) and in 1972 the invention of PDM. The 1980s saw the first PDM polyphase AM transmitter and, in 1987, the invention by Hilmer Swanson of digital amplitude modulation which became known as DX technology.

DX technology is an outstanding example of how the right product can actually create a market which previously did not exist. The US is a large market for broadcast transmitters, but as the market matured there was a tendency for broadcasters to gravitate towards FM broadcasting because of

[1] At the Broadcast Engineering Conference held during the 45th NAB Convention in Las Vegas Lawrence J Cervon was honoured. The Citation reads. "In grateful recognition and appreciation for his 45 years of vision and leadership in the development of high technology broadcasting equipment.' April 14th 1991.

its superior audio response over AM. The introduction of DX changed all that because of its superior AM response and FM-like quality. Now AM broadcasters could achieve FM quality without any of the disadvantages that are associated with FM. Fuelled by a vast home market, Harris DX transmitters scaled powers that grew from the first DX-10, a 10 kW transmitter in 1987, to the DX-25 and DX-50 by 1989, and the DX-100 by 1990. Much of the success of DX is due to the inventiveness of designer Hilmer Swanson, under the leadership of General Manager Thomas Yingst.

In 1990 no one in the broadcast world dreamed that it was possible to take DX solid state technology to power outputs of over 300 kW, except for Swanson and Yingst. Yet these two went on to achieve even more – they broke the 300 kW barrier, then 600 kW, and in 1995 breached the megawatt power level, still using DX all solid state technology. In 1995 DX all solid state technology reached 2 megawatt carrier power, with a capability of 110 per cent modulation on positive peaks.

The Harris Corporation possesses a unique flair for linking high technology with high volume markets, of which DX technology is an excellent example. In the past, high power broadcast transmitters had been hand-built, individually, to order and delivery times were of the order of twelve months or more. The Harris design philosophy changed all that because all the DX transmitter range were built from a number of identical power modules, thus enabling the delivery time for a high power transmitter to be reduced to a few months.

Table 23.1 *Sales of different Harris DX transmitters 1991–1996*

Transmitter	Power (kW)	No. sold	Purchasing country
DX 100	100	24	Algeria, Colombia, Egypt, Korea, Kuwait, Mexico, Oman, Venezuela
Dualled DX 100	200	13	Brazil, Colombia, Ethiopia, Italy, Oman, Spain, USA
DX 100 Frequency-agile version	100	2	Croatia
DX 150	150	1	The Netherlands
DX 200	200	1	Vietnam
DX 300	300	1	Slovenia
DX 300 Frequency-agile version	300	1	Croatia
DX 400	400	1	USA
DX 600	600	10	China
DX 500	500	3	Vietnam
DX 1000	1000	2	VOA (for Thailand and Philippines stations)
DX 2000	2000	2	Qatar, Vietnam

The success of DX was influenced by two other factors; the audio response, as mentioned earlier, indistinguishable from an FM signal, and the overall electrical efficiency of 85 per cent (comparable with only 70 per cent for earlier generations of high power broadcast transmitters). It was the efficiency factor which probably brought the orders that Harris received from VOA and other international broadcasters. This is because the electricity bills for operating very high power transmitters represent a major portion of a transmitter station's running costs and reduction of 10–15 per cent is very significant.

In 1990, in recognition of his work on DX technology, Hilmer Swanson was honoured by the NAB's highest award for engineering achievement. In the five years from 1990 Harris had virtually captured the world's markets for high power transmitters for MW and LW broadcasting. Notwithstanding its huge success with sales of 100 kW SW transmitters, Harris has not attempted to enter high powered SW markets. The reason for this is largely its preferred allocation of engineering resources to products for high volume, rather than low volume, markets.

Chapter 24
Steerable 500 kW rated curtain antenna arrays

Mechanically rotatable curtain antenna arrays for SW broadcasting have become increasingly fashionable since the end of the cold war. Until then, although a few of the major broadcasters had seen the benefits of the rotatable curtain array, the majority had maintained the traditional solution. This was to use a large number of fixed curtains orientated on different target bearings in azimuth. This is a sound concept, but expensive on land requirements; a large SW transmitting station needs anything from 350 to 500 acres (160–200 Ha) of land. However, there are other shortcomings, some of which are of a more serious nature. For example, the idea of having a large antenna farm with up to 40 or 50 fixed antennas will always have the flaw that it will be impossible to direct the output of most or all of the SW transmitters onto the same target zone. Such a deficiency became apparent to major broadcasters VOA and the BBC World Service during the Gulf War in August 1990, when they would have liked to direct most of their output in the broadcast bands to the Gulf region.

On the other hand, those international broadcasters who had invested in having at least one high power, rotatable curtain have never regretted that decision. These broadcasters have included RNI in The Netherlands, ORF in Austria, Radio Vaticana (Vatican City) and a few others. Moreover these international broadcasters continue to use rotatable curtain arrays today.

The first of the big international broadcasters to recognise that target zones may develop rapidly, or move, and also to do something about it was Radio France International. RFI, in conjunction with the state-owned transmission authority TDF, produced the ALLISS concept as a replacement SW transmitter complex to replace the two SW stations at Allouis and Issoudun. The strategy is a network of autonomous transmitting stations, each of 500 kW power, each one of which has the ability to transmit on any desired azimuthal bearing (throughout a full 360 degrees). The output of the entire complex, in this case all 15 transmitters, can be focused onto the same target region if the need arises. Three major broadcasters are now

212

using ALLISS rotatable arrays, RFI, Deutsche Welle and Swiss Radio International, but it is likely that others will follow.

With rotatable curtain arrays coming to prominence, two of the major manufacturers have brought out new models, Telefunken Sendertechnik and TCI.

The TCI rotatable curtain array

The TCI model represents a new concept which addresses the two principal disadvantages of the usual designs for rotatable curtains. One is to do with the fact that rotatable curtains in the past have tended to be used as permanent installations, while the other is their high cost. Cost alone has tended to mask the undoubted technical superiority of the rotatable curtain when compared with the alternative of a number of fixed curtains. TCI's model 611R has design emphasis placed upon an advanced standard of operational performance, while at the same time simplifying its mechanical structure. As a result it is cheaper to manufacture than models based on traditional designs.

Electrically the 611R is a 6 stack high by 2 bay wide curtain with vertical slewing, while the rotatable column provides azimuthal slewing to a full 360 degrees. The 6 high stack gives this antenna its remarkable ability to create a polar diagram pattern with very low departure angles – as low as 1–2 degrees.

The assembly of the curtains consists of two 2 × 6 arrays; a high band array to cover the broadcast bands from 12 to 26.1 MHz, and a low band array to cover the broadcast bands down to 5.9 MHz. These two arrays are mounted in a back-to-back arrangement and separated by an aperiodic reflector screen. The 611R scores over more commonly used 4 stack high × 4 bay wide designs and much improved performance can be seen in the elevation pattern; a choice of departure angles from 25 down to 2–3 degrees.

The physical assembly of the curtains and catenaries uses the well-proven method used in the fixed curtains of the 611 series, in service in considerable numbers with VOA at all its newest transmitter stations, and with others such as the BBC World Service. The only major difference is that the dipoles in the 611R rotable are rigid structures made from aluminum alloy and galvanised steel.

The mechanical design of the TCI rotating structure uses a guyed mast as opposed to the usually massive type of self-supporting structure. This has enabled the designers to simplify and reduce the need for substantial civil works. Moreover, the lightweight mast permits the complete rotatable structure to be capable of being used in a semi-transportable role if required. I believe this feature alone makes the 611R a more suitable antenna than traditional designs for many broadcasters in volatile regions. If the 611R is used in conjunction with SW transmitters housed in transportable

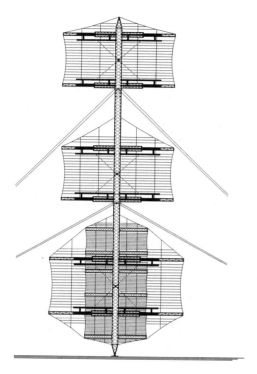

Figure 24.1 *TCI rotatable, vertically slewed SW antenna*

containers, then this concept would enable a complete SW complex of, say, six 611R rotatables and six 500 kW SW transmitters to be relocated to previously prepared alternative locations in the event of war or other change.

Delano SW station: the world's most powerful antenna

Roughly 160 km north of Los Angeles, at the foot of the Sierra Nevada mountains, a giant antenna pours megawatts of power into the sky. The Voice of America's SW station at Delano is an isolated site – bounded to the east by the Nevada desert, including military reservations and the odd nuclear test site; to the south by Death Valley and the Mojave Desert; and by sparsely populated mountains to the north and west, making it ideal for experiments in superpower with electromagnetic energy.

SW transmission is at its best when channelled through directive curtain arrays. Unhampered by serious mechanical problems or the need for tall towers or high-altitude sites, freed from the line-of-sight constraints that beset UHF-TV transmissions, SW sites can take full advantage of electrically long radiators and extended arrays to give a very high concentration of electromagnetic energy.

Figure 24.2 *Two curtain arrays at Delano, CA covering broadcast bands 5.9–26.1 MHz*

Delano transmitting station began its history in 1943, carrying the Voice of America to the Pacific theatre of World War II. Since then it has been in constant service. In 1985, at a critical phase of the cold war, Delano was chosen as the site for the erection of a massive SW curtain antenna – the biggest in the Western world. TCI won the contract to develop, design and build a highly directive curtain array, electronically steerable both

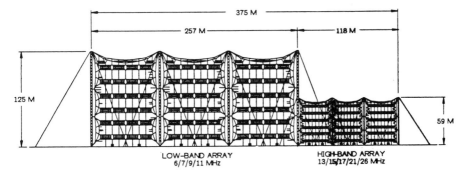

Figure 24.3 *Elevation view of HRS 12/6 array, Delano SW station*

Figure 24.4 *Slewing switch, Delano SW station*

horizontally and in vertical angle above the horizon. Such an antenna would have great flexibility to direct its broadcasts half-way around the world.

Antennas such as these can project a very narrow beam, akin to a sharply focused searchlight. Using optimum frequencies for ionospheric reflection, international broadcasters can aim at any zone or region of the world with confidence that each language broadcast will be most audible in the precise target zone.

When fed from 500 kW transmitters, the gain of the antenna allows it to realise an effective radiated power of a massive 3000 MW – enough, it was thought, to overcome all but the most concentrated Soviet efforts at jamming. The SW curtain array at Delano stands over 125 m high and 375 m long, excluding guys – over 11 acres in area.

Four decades ago, SW broadcasting was more art than science. Today, although the creative element remains, it is a highly developed science. The ionosphere is a constantly changing medium where the strength of a signal can change hugely in a very short time. Transmitter power is an important factor – hence the constant upward trend in broadcast power – but, in the final analysis, the total system performance rests on the antenna.

From the late 1960s, the computer began to play an important role in the design of curtain arrays, making more accurate designs possible. Today's computer technology allows designers to predict very closely the overall performance of a new antenna design.

Apart from the electrical arrangement, mechanical problems also had to be overcome. Swaying and sag, due to wind and thermal expansion, distorted the carefully designed shape and affected electrical performance. TCI addressed this problem using a new composite material, Alumoweld, consisting of a steel wire in an aluminium sheath, used for radiating elements and support wires. The difference in conductivity between copper and aluminium is slight when offset by the advántages of eliminating stretch and corrosion. Its life expectancy is far greater, at upwards of twenty years.

The curtain array consists of two 12 bay wide × six bay high arrays covering all the broadcast bands in the HF spectrum, from 5.5 to 26.1 MHz. With a forward gain of 30 dBi, this array, when fed from three 250 kW transmitters, can produce an effective radiated peak power of 3000 MW.

Each of the 12 × 6 arrays is built from three 4 × 6 arrays, each of which is itself capable of being used as two 2 × 6 arrays. The whole 12 × 6 curtain can be used in a wide variety of modes; the whole curtain may be fed from three 250 kW transmitters, or each 4 × 6 may be used separately, or each 2 × 6 can be excited by its own 500 kW SW transmitter. The wide choice of permutations in width allows VOA to vary the azimuthal coverage, because the azimuthal beam width varies inversely with horizontal aperture.

Changing the configuration to change the radiation pattern requires a matter of seconds. A keyboard in the transmitter building is connected to a computer-controlled switching system at the antenna site, which in turn sets the 99 high-power RF switches to the selected positions.

Delano has certainly the most powerful SW curtain in the Western hemisphere, and possibly the most powerful in the world – although it is possible that the Russians have a larger station, built during the cold war, that they have yet to reveal to the world. The intensity of EM radiation within the site is such that access is strictly forbidden while the station is on air. The installation has also been used in classified US Air Force experiments into heating the ionosphere.

The Telefunken rotatable curtain

There are more Telefunken rotatable curtains in service with more international broadcasters than from any other manufacturer. Table 24.1, extracted from its customer reference lists, shows Telefunken's rotatables to be in service in many countries. If the quantities seem low, it should be explained that until the 1990s – before RFI adopted the policy of using rotatables throughout – it had been the custom for broadcasters to have just one, or at most two, rotatables used purely as a back up so as to be able to take over any of the targeted broadcasts in case of main system failure.

The newest version of Telefunken's rotatables embodies all the company's vast experience in the science of SW broadcasting. Basically this version, the

Table 24.1 *Location of Telefunken rotatable curtains in service*

Broadcaster	Date	Quantity	Power (kW)	Location
Iran IRIB	1963	2	100	Kamalabad
	1992	2	500	Sirjan, Kamalabad
Switzerland PTT	1971	1	500	Sottons
Vatican City State	1976	1	500	Santa M. di Galeria
	1988	1	500	Santa M. di Galeria
Austria ORF	1982	1	500	Moosbrunn
Norway NTA	1982	1	500	Kvitsoy

A 0686/1, uses two wideband curtain arrays comprising full wave dipoles, with a common reflector curtain between them. The HR 4/4/0.5 array covers broadcast bands from 6 to 12 MHz, and the HR 4/4/1 covers bands 13 to 26 MHz. Power handling capacity is 550 kW carrier power with 100 per cent modulation.

There is no doubt that these two new rotatable curtains from TCI and Telefunken will generate interest from those major broadcasters who have

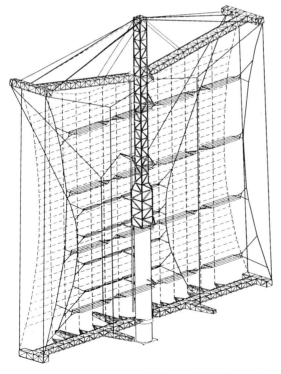

Figure 24.5 *Telefunken 500 kW rotatable SW curtain array*

yet to have the experience of using such a tool. However, no direct comparison between the two is possible because they are intended for different markets. The lightweight version from TCI with its ease of relocation to another site, will have appeal for many countries in the Middle East and elsewhere. On the other hand, the robustly engineered model from Telefunken, which embodies the company's unrivalled experience in antenna design, mechanical stress control and design for long life, will make the A 0686/1 attractive to international broadcasters like the BBC World Service, as well as certain Middle East broadcasters, such as IRIB Iran, Voice of Turkey and others, who make long-term investments in foreign service broadcasting.

Chapter 25
Profile of the tube manufacturing industry

There have been two events in history that brought about the massive decline in world markets for thermionic tubes. The first was the invention of the transistor in 1947 by Shockley and co-workers at the AT & T Bell Telephone Laboratories; the second, a direct consequence of the first, was the capture less than two decades later of world radio markets in the domestic sector by transistor radio sets. It was the transistor, by that time in full scale mass production, which made it possible to produce the highly compact battery-operated portable receiver. By the early 1960s a mass market had been created in Asia, the Middle East and elsewhere around the world. The globalisation of this market heralded the decline of mass production and sales of tubes.

Back in the 1940s the production of tubes was at its peak. World War II generated a massive market for tubes of all types, from small signal amplifiers to power grid tubes for use in communications, radar and radio broadcasting. America became the powerhouse to drive the Allied war effort and production increased dramatically at the plants of RCA, Westinghouse, General Electric and Machlett. The microwave power tube industry also boomed during this period, with companies such as Raytheon, Sperry, Litton and others, including Eimac, which at its peak employed 3000 workers.

The end of World War II saw a drop in tube demand whose speed matched that of its rise from 1940, and war surplus tubes came onto the market. However, as demand began to fall from mass markets for domestic products, so the technology of tubes for professional applications began to rise. Research and development costs were increasing but without the benefit of a mass market to sustain the increased amount of research: the tube industry had moved to a state of diminishing return on capital investment. In the post war years the major tube companies that survived and prospered in the US were Eimac, Machlett and RCA, while in Europe the major companies were ABB, Philips, Siemens, Thomson and EEV in Britain. As the market for higher powered tubes for power grid applications in

220

SW transmitters began to expand the major transmitter manufacturers in the US, such as Hughes, Continental and Harris, tended to design around US-built tubes, while in Europe the transmitter manufacturers favoured the use of European tubes from Thomson, ABB and EEV. Behind the iron curtain the major tube manufacturer was Svetlana in St Petersburg. Its high power tubes were used in high power transmitters and jammers throughout the USSR and other Eastern bloc countries.

In 1993 some contraction of the tube manufacturing industry began to take place. Thomson Tubes acquired the Tube Division of ABB and, along with it, tube manufacturing plants in Switzerland and Poland to supplement existing plants and R & D centres in Thonon, near Geneva, and three other plants at Meudon-La Forêt, Saint Egrève and Moirans, all in France. In 1995 Thomson Tubes also acquired the tube manufacturing business of Siemens in Germany, thus creating Thomson Tubes Electroniques, the largest manufacturer of tubes in the world.

In June 1995, the California-based Varian Associates, which had acquired Eimac in the 1960s, decided to exit the tube industry for good. It sold off its interests to Leonard Green & Partners, who continued to produce the Eimac product line. Also in the US, the RCA tube division was acquired by Burle. Today there also continues to be a number of smaller companies which specialise in rebuilding power grid tubes, notable of which is the California-based company Econco, started by ex-Eimac employees.

From an industry which began a hundred years ago to serve the age of wireless broadcasting, and which started with hand-built, primitive, crude tubes, has developed one of the highest of technologies. It is a truly multidisciplinary industry which has to continue making a vast expenditure in R & D in applied physics, chemistry (thermodynamics), metallurgy, electronic emission, and the generation of huge amounts of radio frequency power. Research and development costs are several times greater than for the transmitter manufacturing industry, for instance; a fact which makes it all the more difficult for the smaller manufacturers to survive.

The tube technology developed by Svetlana in St Petersburg during World War II and the cold war and since then, is now available in the West in the form of plug-in compatible replacement tubes for Western broadcast transmitters. Though presently available for low and medium power applications, it can be expected that this Russian tube manufacturer will increase the power levels of its tubes thereby making the option of using Russian-built tubes available to Western broadcasters in their high power broadcast transmitters. The experience and ability to build very high power tubes is clear from the number of transmitters and jammers that operated in the cold war.

The history of Thomson Tubes

The original company owed its existence to Elihu Thomson, born in 1853, the son of Scots parents who had emigrated to America. At the age of 25, while working at Philadelphia High School, he discovered (albeit accidentally) the existence of a new phenomenon – electromagnetic energy. This discovery was not widely publicised but it nevertheless laid the groundwork on which Marconi and Hertz expanded. In so doing, Thomson became one of America's most eminent scientists and went on to make further contributions in the electrical sciences. Yet, notwithstanding his many great achievements, his greatest legacy to the scientific world is perhaps the company he formed with his partner Houston, which eventually led to Thomson–Houston companies being formed in the US, Britain and France.

Today that company is Thomson SA, with more than 110 000 employees around the world. Involved with electrical engineering from its early origins, the company's activities are still centred in the electronics sector, in both consumer and professional markets. From its early days the company has been involved with the manufacture of tubes and transmitters for radio broadcasting. The company then known as Compagnie Français Thomson–Houston (CFTH) made a major breakthrough in tube technology, the Vapotron, and in 1955 brought the world's first 100 kW Vapotron-fitted SW transmitter into service at Leopoldville in the Belgian Congo (now Zaire) in Africa.

Figure 25.1 *Thomson Hypervatron Tetrode high efficiency tube, rated at 550 kW in SW serivce (650 kW for LW)*

The company's second major breakthrough was the pyrolitic grid, after which it went on to become the acknowledged world leader in tube technology. By the late 1970s Thomson dominated the world markets for high power radio tubes. In Africa alone Thomson tubes are in service in more than 30 countries. In 1988, in keeping with the Tube Division's achievements, it became a separate company within the overall Thomson group – Thomson Tubes Electroniqies (TTE), an independent operating company of Thomson-CSF. Today it is the largest tube manufacturer in the world, employing a workforce of 2300, of whom a high proportion are highly skilled in a number of related disciplines. TTE's activities in 1996 amounted to 1.5 billion French Francs, with some 16 per cent of that being invested in research and development into tubes for the next generation.

When the thermionic vacuum tube was invented its development stemmed from the need for a device for radio broadcasting. Today tubes have found applications in a wide number of applications and these range in physical size from the giant superpower pulsed klystrons, seven metres tall, down to 'one-inch' camera tubes. About two thirds of TTE's turnover is in civilian markets for telecommunications, space communications, radio and TV broadcasting, medicine, industry and science, and one third from military markets.

Tubes with hot cathodes are going to be used for a few more decades simply because there is no comparable device to replace the tube when it comes to generating large amounts of radio frequency power, particularly at frequencies higher than a few megahertz. In the meantime, solid state will continue to make further inroads in those areas where the technology is a viable alternative – even if that necessitates the use of very large numbers so as to make up the required power handling capacity. The tube industry is shrinking in size but not in its technology; research and development goes on continually. In the broadcast world the last bastions for tubes are high power FM transmitters, in UHF-TV, HDTV and particularly in SW, where the classical design of the high power RF stage is set to survive for many more decades.

Chapter 26
The future is digital

The principal international broadcasters of Europe, the BBC World Service, Deutsche Welle and Radio France International, are unanimous in their belief that SW broadcasting in the HF bands will continue to be the main gateway for international broadcasting for several more decades. The political advantages are such that no other technology, existing or foreseen, will have the ability to reach out directly to listeners in any part of the world without a third party intervention. More than ever before, international broadcasters must retain their unique ability to reach the hundreds of millions of poor and oppressed, in places where truth and news do not circulate easily.

Yet, at the same time, it has to be recognised that the future lies with digital rather than AM. Digital technology is advancing fast and broadcasters recognise this. AM has served the world's broadcasters for more than sixty years but now is the time to plan for the next sixty. For all its versatility, it has to be admitted that SW AM is far from a perfect medium;

> 'The poor transmission quality that is associated with AM transmissions in the long, medium and particularly the shortwave bands arises mainly from the nature of the modulation method, and less so from the frequency band. By switching to a form of digital modulating modulation in these AM wavebands, the quality of transmission will be greatly improved.' [13]

Digital transmission is not an untried, unproven technology. Broadcasting digital sound for distribution over satellite links began in 1985, and since 1991 VOA, for example, has been distributing 40 audio channels around the world on its Americas and Caribbean digital audio service via Intelsat 601 and Eutelsat 2F1.

Digital transmission offers several advantages, the main of which are that the transmitted signal is more robust, extraneous noise is virtually eliminated and it delivers a better quality of signal to the end user. A further key advantage is a lowering of transmission costs because a smaller size of satellite dish can be used than for analogue earth stations. VOA now spends on digital only one third of what it previously spent on its analogue satellite earth stations. If digital audio is used in the AM wavebands instead of AM it results in programme quality that is far superior to AM at its very best, able to deliver full CD quality if required, with the additional benefit of transmitting stations running more efficiently and with lower amounts of kilowatt output power.

A further urgent reason for introducing digital broadcasting in the AM wavebands has come about since the end of the cold war. FM radio stations have appeared by the hundred in the former Soviet Union and Eastern European countries. Broadcasters such as VOA and the BBC-WS were quick to exploit the possibilities of reaching larger audiences by re-broadcasting over these FM stations. The downside to this, however, was that several hundred million SW listeners in these countries heard FM quality for the first time, with the effect of making AM broadcasts less attractive.

At the same time it has to be recognised that SW broadcasting remains the tool most suitable for projecting foreign policy, news and views to many regions of the world, particularly Africa, Asia and the Far East, and it is important that those listening audiences should not suffer any decline. The ideal medium for international broadcasting in the future is to retain SW as the means of delivery, but with simulcast broadcasting – transmitting the same programme in both analogue and digital formats. Concurrently, economically priced digital/conventional receivers should be promoted. This would enable those listeners with digital SW receivers to hear superior sound quality programmes, while at the same time not depriving those listeners with ordinary AM receivers. Eventually, over a period of time it should be possible to cease the analogue sound programme, by which time receivers with the chips to permit digital reception will be available at low cost.

That the AM wavebands possess important propagation qualities is shown by the upsurge in investment in high power and even superpower MW and LW transmitters by several broadcasters in the mid 1990s. Broadcasting in the long and medium wavebands projects a very stable signal over very long distances, with none of the handicaps and disadvantages of FM broadcasting such as adjacent channel interference, fading or signal loss. Therefore considerable advantages would result from the employment of digital modulation not just in the SW bands but across the AM spectrum from LW to MW. The implementation of digital broadcasting could retain the advantages associated with AM broadcasting but without the disadvantages of analogue sound AM.

In 1996 the ITU established a working party to study a proposal for implementing digital broadcasting systems in all AM wavebands. The resulting document 10A/15 defined the selection criteria for digital modulation, made proposals for different systems and outlined advantages for such an introduction. It concluded that AM analogue transmission could be replaced by digital transmission without any difficulty. Its introduction would result in the advantages listed below:

- better quality of audio;
- immunity from echo interference;
- longer transmission range for the same power;
- avoidance of fading and interference in AM analogue broadcasting;
- lower transmission power to achieve the same coverage of AM;
- superior all-round performance, especially in SW broadcasting.

Digital sound broadcasting comes in different formats ranging from full CD quality with stereo sound, near-CD stereo, near-CD mono, FM stereo, FM mono and down to ordinary AM quality depending on the amount of digital compression in the digital modulation process. The ITU report further concludes that a digital receiver can be easily designed for demodulating the conventional AM signal as well, and that switching between analogue and digital modes could be carried out automatically resulting in improved audio quality.

The ITU's Digital Radio Mondiale (DRM) system is under serious discussion amongst leading international broadcasters such as the BBC-WS, DW (Deutsche Welle), RFI, NHK (Nippon Hoso Kyokia) and VOA. Two European manufacturers have developed systems that permit simultaneous transmission of a digital audio signal and a conventional AM signal within the passband of existing SW transmitters. The only additional equipment needed at the transmitter is a digital modulator.

The enormous advantages that would be achieved by implementing digital broadcasting on the short waves alone more than justify whatever further research is necessary. These advantages are political and technical, as well as being economically sound, and because international broadcasting in the HF spectrum is well regulated, the same regulatory bodies can stay in place without the need for any more frequency allocations.

The economic advantages are truly great because there are more than 200 high power SW installations around the world, many of which are modern, all of which could be converted to transmit a digital format and the only additional costs envisaged would be the addition of a digital modulator. Multi-transmitter high power SW sites are an expensive investment (typically $70–100 million) so no broadcaster can afford to consign such facilities to the scrapyard in favour of unproven alternative methods such as satellite delivery systems. Digital delivery offers a viable alternative.

Of course it also makes good sense to introduce digital format in the other AM wavebands, LW and MW, with their unique qualities, where there is now increasing interest from many countries.

Progress in the digital field

The first transmitter manufacturer to devise a workable method for a SW digital service was Thomcast. This system, called Skywave 2000, was first demonstrated at the Third Montreux Radio Symposium in 1995 where Thomcast showed it was both feasible and practical to transmit a conventional analogue single sideband (SSB) signal within the lower sideband (LSB), with a simultaneous digital transmission within the upper sideband (USB).

Concurrently, and independently, Deutsche Telekom AG had started some tests with digital transmissions in the medium waveband in October 1994 from Berlin Kopenick, using a frequency of 810 kHz and 1 kW transmitter power. These tests yielded some very satisfactory results, and in spring 1996 Deutsche Telekom, Europe's largest telecommunications authority, began making experimental transmissions on SW from its high power SW station at Julich on a frequency of 5910 kHz.

On 25th April 1997 the digital revolution was taken a significant step further when the T^2M system was announced, a joint research and development programme by Telefunken and Deutsche Telekom. T^2M is different from the Thomcast Skywave 2000 system offering possibilities for:

- transmission in all wavebands (LW, MW and SW);
- spectrum transmission of analogue in double sideband;
- simultaneous transmission of a digital signal;
- possibility of 1 or 2 digital programme channels;
- CD quality when combining both digital channels;
- no mutual interference between AM and digital.

T^2M also offers reduced effects from fading and less difference between daytime and nighttime reception in the medium waveband. Furthermore, lower carrier powers can be used to cover the same target areas, compared with ordinary AM.

The Thomcast Skywave 2000 is a logical development from the World Administrative Radio Conference (WARC) plan to introduce single sideband transmission to the HF broadcast bands by the year 2015. Skywave 2000 is not, however, intended for use in all AM wavebands because it relies upon SSB transmission, although, there is no reason why it could not be used in all AM wavebands if SSB were to be used on LW and MW.

On the other hand, T^2M is suitable right now for all AM wavebands because it does not depend on the use of SSB transmission. Moreover, because the Telefunken method will support the transmission of two digital channels, which can be combined to produce stereo sound, the T^2M system is capable of delivering full CD quality.

Common to both systems is the need for a new generation of AM receivers with the additional chips to enable reception of digital transmissions. Until

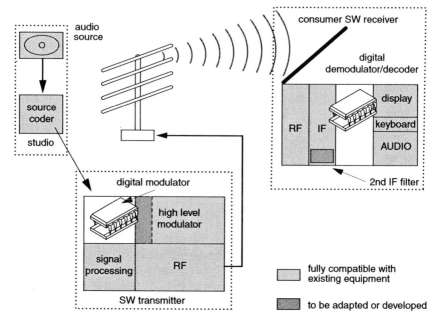

Figure 26.1 *Thomcast Skywave 2000 system*

such receivers become available in cost-effective large-scale production, digital broadcasting will not advance rapidly. As with all emergent technologies, initially the cost of such receivers is high, and only mass demand will bring down costs within the reach of all.

AM band characteristics

The AM wavebands include long wave, medium wave and the broadcast bands in the HF spectrum commonly called shortwave. Each of these designated bands possesses unique transmission qualities:

- *Long wave* 150–285 kHz, channel width 9 kHz. It exhibits stable transmission qualities, is generally immune from fading with little or no interference from sky-wave (ionospheric reflection) and is ideal for long range radio broadcasting using high power, from 500 to 2000 kW.
- *Medium wave* 525–1605 kHz, channel width 9 kHz or 10 kHz. During hours of daylight propagation qualities are as for long wave but with slightly reduced range. Propagation during the hours of darkness is considerably extended – even to several thousand km, using high power and directional antennas. Selective fading can occur at the edges of the daytime range. The higher the operating frequency, the better the nighttime extended coverage.

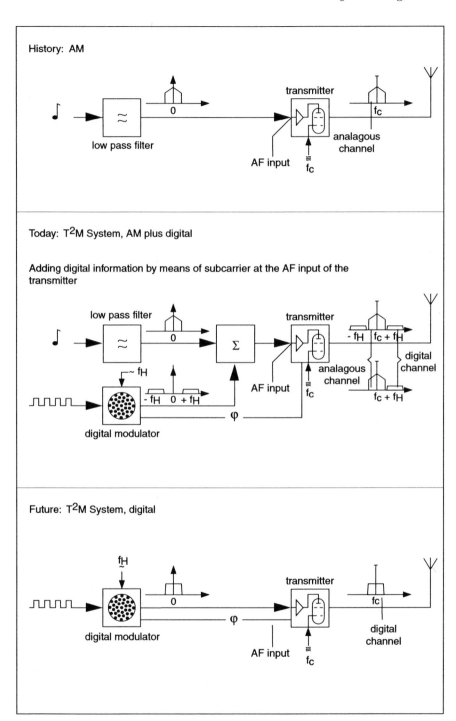

Figure 26.2 *Telefunken T²M multi-cast system*

- *Shortwave* 3.3–26.1 MHz, channel width 10 kHz for DSB and 5 kHz for SSB. All propagation is essentially by sky-wave. The ideal propagation path is 3500–4500 km with a one-hop ionospheric path. Distances of 3500–18 000 km are possible with a multi-hop propagation path. The ionosphere is a dispersive propagation medium, subject to vagaries and sudden changes in signal strength and is affected by multimodes, with each mode presenting a particular group delay, amplitude, polarisation and doppler frequency shifts.

The introduction of digital broadcasting in all these AM wavebands, but especially in the case of MW and SW, would mitigate to a very large extent the unwanted behaviour arising from the vagaries of the ionosphere.

Recent upsurge in investment in LW and MW broadcasting

In contradiction of the forecasts made by some media experts in the late 1980s predicting the decline of LW and MW broadcasting brought about by FM and the promise of DAB, my own evidence indicates not only that AM broadcasting is as popular as it was in the 1960s, but that there is an upsurge taking place. When this upsurge takes the form of investment in 1000 kW and even 2–3 megawatt installations, we can be sure that others also believe this.

In Western Europe alone, three broadcasters have either acquired or ordered two megawatt AM transmitters; Europe No. 1 Felsberg, Radio Tele-Luxembourg (RTL), and most recently Delta 171 of the Netherlands. Such sales are only part of the global picture because Harris has sold a much larger number of one and two megawatt transmitters to the Middle East, SE Asia and China. Sales of high power AM transmitters began to increase from 1993 onwards, and according to an analysis of global sales, 38 megawatts of AM power, from 300 kW upwards, have been sold. If account is also taken of sales of transmitters with less than 300 kW carrier power, the results are even more impressive. Table 26.1 shows the total sales in kilo-watts of transmitters of 300 kW and over between 1990 and 1994.

Table 26.1 *World sales of transmitters of over 300 kW power, 1990–1994*

Company	MW (kW)	LW (kW)	Total (kW)
Thomcast, France	8400	5500	13 900
Harris, USA	16 400	400	16 800
Continental, USA	1200	0	1200
Telefunken, Germany	6300	0	6300
Marconi, UK	1200	0	1200

Note: Excluded from the above analysis is a recent order from DELTA 171 for a 2 mW LW station, for which the prime contractor is Telefunken Sendertechnik.

Chapter 27
The future for international broadcasting in the HF spectrum

When I began writing this book, Volume 2 of *History of international broadcasting*, it was 1992, the cold war had ended and the Gulf War had taken place. It is an evolving history and drafts of some chapters were being written as events unfolded. Thus, although there have been some significant developments since 1992, I have tried to put these in context. For example, in discussing the restructuring of the US Government's international broadcasting assets, I have taken care to show how this evolved, and to list the transmitter networks of Voice of America as it existed before the rationalised network of the International Broadcasting Bureau (IBB) came into being.

Similarly I have devoted much space to describing the operational and technical status of the RFE/RL international broadcasting network as it existed in the closing years of the cold war. RFE/RL was a unique organisation in many ways – its covert nature as a privateer made it immune to bureaucratic control and questions from Congress. Above all, it was a uniquely structured propaganda broadcasting network which fulfilled the missions it had been set, this success partly made possible by some highly sophisticated engineering in transmitters and high gain curtain arrays.

I believe it is right and proper that the achievements of RFE/RL should be recorded, and details of its unique network maintained for future historians. Governments have a way of rewriting history, especially when it might be politically expedient. There has been much official publicity about the achievements of VOA (some from VOA itself) in the bloodless revolution which led to the overthrow of communism, but little mention of what RFE/RL achieved. Similarly, some RFE/RL transmitter stations have been dismantled, while others have been stripped of transmitters and antennas. For these reasons I would have liked to write and preserve more about RFE/RL than I already have, but space precludes me from so doing.

The success that was achieved during the cold war by the US Government's international broadcasting arms is best summed up by the President's Task Force on International Broadcasting when its Chairman said in December 1991, 'Money spent by the US taxpayer on funding the operations of VOA and RFE/RL was one of the best investments that America had made.'

From that public acknowledgement it was almost a foregone conclusion that the same weapon would be used to fight communism in Asia. Now Radio Free Asia has been in operation for two years. Presently these broadcasts, which are reaching citizens in China, are on nothing like the scale of the cold war with the Soviet empire. Nevertheless, according to reports coming out of Washington DC, good headway is being achieved. One popular programme is 'Literature you can't hear', which features readings from books which are banned in the People's Republic. A prominent commentator is Wang Dan, the 1989 Tiananmen Square student protest leader, recently released from prison. Two of RFA's most famous listeners are said to be the Dalai Lama, exiled spiritual leader of Tibet, and Aung San Suu Kyi, opposition leader in Myanmar.

Evidently the success being achieved with these broadcasts to China has created enthusiasm in US government circles to establish other US-sponsored 'Radio Free' broadcasts. A 'Radio Free Iraq' is likely to come on line, and some proponents believe it could have a better chance of success than military power in removing Saddam Hussein from office. Other proposals may include tentative ideas for a 'Radio Free Iran' but this is uncertain. For the foreseeable future, the priorities are believed to be RFA and RF Iraq.

As is the case with almost all international propaganda broadcasting, it is dissidents who have been co-opted to carry out propaganda broadcasting to their former countrymen, exposing things like corruption among high officials. More often than not, these dissidents are not broadcasting from within the targeted country, nor even elsewhere in Asia, but from the relative comfort of a broadcasting studio in Washington DC. For example, in November 1997 the Chinese dissident Wei Jingheng, released from prison on health grounds, gave his radio interview to VOA, from Washington.

With all propaganda broadcasting the actual number of people who hear the broadcast is not necessarily crucial to success; although it is useful for broadcasters to know approximate figures for listener numbers, a low count is not necessarily an indication of failure. This is because for every sympathiser, dissident or SW enthusiast who hears an account of corruption in high places or abuse of privilege, the news report will be spread by word of mouth and through underground newspapers. Indeed, it may even be the case that if only a few hear the actual broadcast, this imparts a certain exclusivity to the story, encouraging others to say that they too heard the broadcasts. This was proved during World War II when both Germany and Britain achieved large measures of success with propaganda broadcasting.

Broadcasters such as the BBC World Service do not appear to be too concerned about the loss of listeners in Western Europe. The real targets for SW are Africa, the Middle East, South Asia and the Asia Pacific region. An examination of Table 6.2 (Chapter 6) for the BBC World Service listening audiences emphasises this fact. Of an audited figure of over 130 million, Africa and the Middle East account for 42.5 million, whilst Asia and the Pacific rim account for 59.5 million; 102 million of its listeners are in these regions of the world. The remaining 28 million are in the Americas, (Canada, USA, Central and South America, 8 million), Europe (including Central and Eastern, 12.5 million), and the former Soviet Union and SE Asia (7.5 million).

The decline in world SW listening audiences

Since the end of the cold war and the overthrow of Soviet communism there has been a gradual decline in world audiences for SW. Specifically the decline has been most noticeable in the republics of the former Soviet Union, the Baltic States and the former Eastern bloc countries. The decline can be attributed to three main factors. First, the overthrow of communism diminished the need for many of the listeners to tune into Western broadcasts from the BBC, VOA and RFE/RL. Second, FM broadcasting was gradually introduced into those countries where SW was a staple, along with broadcasting in the long and medium wave bands. Third, there was the growth in pluralism, a more stratified society and the emergence of a comparatively free media in some countries.

All of these factors have contributed to a loss in SW listening audiences in these regions of the world, but in this respect SW is to a certain extent the victim of its own success. Western broadcasters such as the BBC and Deutsche Welle concluded agreements at the end of the cold war for countries such as the Czech and Slovak republics to take up their SW broadcasts and relay them at no charge over local FM and AM stations. This was a way for these Western broadcasters to increase their listening audiences; the BBC, for instance, estimates it picks up another 30 million listeners this way. The danger, however, is a concomitant long-term loss of SW listening audiences who have been weaned off SW.

But in addition to a declining audience in Eastern Europe and the former Soviet republics, we are seeing a steady decline in SW listening in Western Europe too. This situation has come about for two reasons. Business travellers who would at one time take their SW portable receiver on business trips overseas can now see or hear international news from CNN and other international news agencies by direct satellite broadcasting. The second contributory factor to the decline, as discussed, is competition from many other activities – leisure time in Western societies is catered for by many other media.

However, the real threat to SW listening in the West comes about because it is not a market-driven business and as a result has a low profile. Many people still think of SW as a hobby for radio amateurs, or else an outdated technology. The truth is quite the opposite; SW is a highly developed science and the technology has never been better. Modern SW transmitters use very high power, many are of the 500 kW class, and can deliver robust, high grade, stable signals to the other side of the world with equal facility.

Throughout the past one hundred years or so, youth has demonstrated that it is the first to be attracted to any new technology. When SW made its debut nearly seventy years ago it was embraced by younger people, and today youth is still demonstrating its fascination with new technologies like the Internet. Such technologies are sold through massive co-ordinated advertising. The advantages of digital television are already being promoted. The fact is that many of the advantages are with the equipment manufacturers, broadcasters and programme makers, yet, despite this, digital television is likely to be a marketing success.

It is perhaps too elementary to point out that without a listening audience, the SW market will wither. From this equation it will be obvious that the international broadcasters should be looking for ways to increase SW listening audiences – where better to start than in their own country? But apart from Radio Netherlands International, which is trying to popularise SW listening among the Dutch public, most major broadcasters do little to popularise SW at home; indeed, rather the opposite. It is not all that long ago when a British government minister described the World Service as 'one of Britain's best kept secrets'.

Over the past fifty years, many governments have adopted a cautious attitude to foreign media because of the potential threat to internal stability. These include foreign printed media, satellite television, and also SW. Neither the US nor the British governments reveal estimated figures for SW listening at home, but during the cold war they went to great trouble to find out the numbers within the USSR and Eastern Europe.

But the cold war is over, and there is little danger of Western societies being subverted by foreign propaganda. Yet most Western governments do not publicise the merits of SW at home. A frequent excuse is lack of funds, but this is surely not the case. The BBC World Service, for instance, could probably be promoted through the BBC national networks at little cost. In some countries governments go beyond this lack of encouragement for SW at home, and even actively discourage it. Voice of America is forbidden by its charter from sending the same programmes to US citizens at home as abroad, and it is quite difficult to pick up the BBC World Service in Britain.

Of the big international broadcasters it is the BBC World Service that has the highest profile and largest audience worldwide with an audience of approximately 30–40 million more listeners than its nearest competitors, Voice of America and Radio France International. The best opportunity to show the British public what this medium, SW, can do came in 1990 after the collapse of Soviet communism. This was a spectacular achievement by the West in which Britain's BBC World Service played a key part, but at no time then or since has the British Foreign Office given it the credit that is due.

References

1 US President's Task Force on US Government International Broadcasting, 1991
2 WOOD, J.: 'History of international broadcasting' (Peter Peregrinus (IEE) 1994)
3 TUSA, J.: 'A world in your ear' (Broadside Books 1992)
4 WALKER, A.: 'Skyful of freedom' (Broadside Books 1992)
5 WOOD, J.: 'Broadcasting the voice of America!' *International Broadcasting*, July 1987
6 WOOD, J.: 'VOA: Born during the dark days of war' *Radio World*, January 1995
7 KERSHNER, S.W.: 'High performance antennas for new VOA stations' *IEEE Trans. Broadcasting* **34**(2), June 1982
8 Radio Canada International 51st Anniversary Publication, 1996
9 KOOMANS, N.: 'Single-side-band telephony applied to the radio link between The Netherlands and The Netherlands East Indes' *Proc. IRE*, 1938, **26**(2), pp. 182–206
10 SMITH, N.S.: 'High freqency broadcasting in Australia' *Proc. IRE*, 1948, **9**(10), pp. 4–20
11 STEELE, A.: 'Loud let it ring' (Pacific Press 1997)
12 BARNOUW, E.: 'A history of broadcasting in the US'
    Volume 1: 'A tower in Babel' (1966)
    Volume 2: 'The golden web 1933-53' (1968)
    Volume 3: 'The image maker 1953-70' (1970) (OUP, New York)
13 Rudolph, Prof Dr Ing. D. (Deutsche Telekom Berlin Engineering College): 'Principles and proposals for digital radio broadcasting using long, medium and short waves', lecture delivered in Mannheim, 1995

Bibliography

BARNOUW, E.: 'A history of broadcasting in the US'
  Volume 1: 'A tower in Babel' (1966)
  Volume 2: 'The golden web 1933–53' (1968)
  Volume 3: 'The image maker 1953–70' (1970) (OUP, New York)
BOBBETT. D.G.: 'WRTH World Radio TV Handbook 1999' (Watson-Guptill Publications, 1999)
CALVOCORESSI, P.: 'World politics since 1945' (Addison Wesley Longman Higher Education, 1996)
HALLIDAY, F.: 'The making of the cold war'(Verso & NLB Pubications, London, 1983)
MOSELY, L.: 'The Dulles family' (Hodder & Stoughton, London, 1978)
NIXON, R.: 'The real war' (Sedgwick & Jackson, London, 1980)
WOOD, J.: 'History of international broadcasting' (Peter Peregrinus Ltd., Stevenage, 1992)

The reader is also directed to a series of articles by the author in *Radio World* (International Edition), *International Broadcasting* and the *IEE Review* as well as other technical periodicals.

Appendix 1

Radio Free Europe/Radio Liberty

RFE/RL Inc. had its administrative headquarters in Munich, Germany. Housed within the building complex were its news-gathering and evaluation activities, political research groups, programme production and broadcast studios, together with information services, publications and engineering, administration, planning and maintenance facilities.

Major offices were also located in Washington DC and in New York. These provided liaison with the Board for International Broadcasting as well as other US government agencies and provided engineering, technical and purchasing support. Studio facilities at both of these locations provided US-originated programme material. A receiving station at Schleissheim near Munich monitored Eastern and Western news and information services which were then evaluated by the Munich personnel.

Finished programme material, either live or pre-taped, was then distributed over satellite and leased lines, or by radio relay back-up, to transmitting stations located at Biblis, Holzkirchen and Lampertheim in Germany, at Gloria, Portugal, and Playa de Pals, Spain for transmission on MW and SW frequencies to target areas in Eastern Europe and the Soviet Union. Additionally, as the decade of the 1990s opened up, RFE/RL initiated an exciting new chapter in its history. This was the feeding of programme material directly from its Munich studios into former Eastern bloc target countries for broadcasting locally on MW transmitters.

The locations of the major European RFE/RL facilities are shown in Figure A1.1. A block diagram (Figure A1.2) shows the programme distribution network as it was in 1991. Geographical separation of transmitter sites from programme production centres, and the attendant need for elaborate intercommunication facilities over very long distances, may at first seem unusual to a person unfamiliar with SW broadcasting, and more familiar with a low power FM radio station which might have all is facilities located in one building. However, SW broadcasting in the HF spectrum is an entirely different activity.

Figure A1.1 *Location of RFE/RL transmitting stations in Europe*

Siting is the key to the effectiveness of a SW transmitter site and RFE/RL gave special attention to this. SW broadcasting produces maximum audibility when the target area is 2500–4000 km distant, using a one-hop ionospheric transmission path. When target areas have been established, the hours of operation and broadcast bands have been calculated, and the necessary licence to operate the station obtained, a search for a suitable site may begin. This is a highly technical project because it calls for a site which is remote from densely settled areas, with good electrical conductivity and with no obstructions in the target azimuth greater than one degree of elevation. To facilitate this the transmission site is best located on flat terrain. The employment of departure angles down to grazing i.e. 2 degrees, minimises signal attenuation in the ionosphere. Different frequencies and bands can be selected according to times of day to achieve optimum performance. Additionally, the further south a northern hemisphere station is located, the longer the higher frequencies are useable during winter months when hours of darkness are extended.

These are just a few of the technical factors which play a part in the siting of SW transmitter facilities, and explain why major international SW broadcasters have located their transmitter stations in particular places around the world. Other considerations range from the availability of telecommunications, electrical power and water, to good access by road and

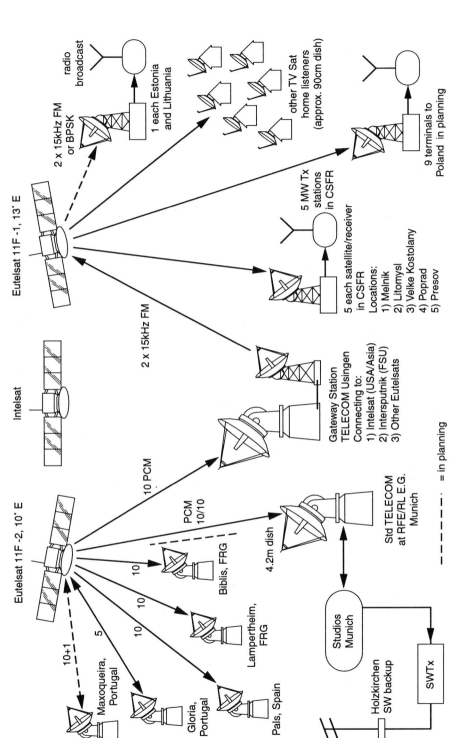

Figure A1.2 *Block diagram of RFE/RL facilities and programme delivery*

rail, but, above all, it is vital that the transmitter stations are located in host countries that have political stability so that long-term agreements can be concluded.

It is far from easy to find countries and sites which satisfy all the legal, technical, logistical, economic and geopolitical factors. RFE/RL was extremely fortunate to be searching for sites at a time when Germany was dependent upon America for very many things, and when some sites with excellent characteristics (such as ex-Luftwaffe airfields) were readily available. Transmitter stations in Germany, Portugal and Spain were acquired when construction costs and land prices were low. The value of these RFE/RL sites in present day prices could make their development today impossible.

Technical site descriptions

Schleissheim receiving station

This station was located at 48.14.3N × 11.33.3E, north of Munich. Its function was content monitoring of broadcasts from sources from the capitals of Eastern Europe, through the Soviet Union and as far as Peking. Signals were received and then relayed over telephone circuits to Munich where trained translators converted them to print for editorial use. As well as speech, the monitored broadcasts included reception to RTTY (radio teletype).

The routeing of transmissions from the receivers at Schleissheim to translators in Munich was accomplished by a sophisticated automatic, programmable cross-bar system which permitted outputs to be automatically routed to the appropriate telephone line according to a pre-arranged schedule. The system also permitted connection to tape recorders for storage, for later transmission to the translators. Reception of news was from a number of receivers and a variety of directional antennas beamed on the source of the broadcasts. The antenna site covered an area of over 235 acres (95 Ha).

Continuous wave (CW) cueing signals from Portugal for the operation of the Munich–Holzkirchen–Maxoqueira–Gloria programme relay were picked up at the Schleissheim station and sent over telephone lines to the Holzkirchen transmitting station. All these services required a considerable number of telephone circuits between the Schleissheim station and the Munich studios.

Munich studios and offices

Located in Munich at Oettingstrasse 67, Am English Garten, these facilities were housed in one large building. This was the nerve centre of operations;

the communications and broadcast circuits that terminated here brought most of the information that formed the basis of RFE/RL broadcasts. Programmes were produced either for live broadcasting or for delayed broadcast from tape. Either procedure could be accommodated by a computer-controlled automated system. This system could start and stop pre-recorded tapes, cue announcers and feed a properly sequenced programme to the transmitters. The total production capability of the studios was extensive and easily capable of accommodating all language broadcasts from RFE/RL transmitters on a simultaneous basis.

Extensive communication circuits linked Munich with the Washington and New York offices, the various receiving and transmitting sites, and many other miscellaneous locations. The function of these circuits is best understood by a description of these centres; New York and Washington were connected by three 5 kHz programme grade simplex circuits and two 128 kbit/s to the VOA operations centre at Ismaning, Germany, from where they ran to the Munich studios over 7.5 kHz analogue lines.

US operations

RFE/RL's US operations are located in the Washington office, the New York Programming Centre and the NCA US Bureau. Its technical facilities in the US include studios with remote control and broadcast equipment, telephone systems and programme transmission systems. All are managed by the Broadcast Operations Department, US (Bod-US).

RFE/RL New York Programming Centre (NYPC)

The NYPC was located at 1775 Broadway in New York City. The facility housed the production department and technical areas, RFE/RL editors, Engineering Division frequency management, broadcast operations, NYPC administrative offices, information services and library. Technical areas included four studios and control rooms, a tape dubbing facility, a master control with two announcer booths, four editing rooms and various other facilities and services. From this location RFE/RL broadcast departments regularly produced thirteen programmes in Soviet and East European languages.

Technical monitoring

The frequencies used by RFE/RL in the HF broadcast bands were subject to agreements with certain major Western broadcasters in the UK, Canada and US, as well as the host countries Germany, Spain and Portugal, because broadcasts from these also affected RFE/RL. It became necessary from time to time to justify channel usage to friendly broadcasters and to make strategic decisions with respect to others. For these reasons, and some others, RFE/RL set up regular monitoring offices as near as possible to the edge of the target areas. These monitor stations were equipped with field strength measuring equipment and a number of radio receivers. They noted the characteristics of both wanted and unwanted signals, the occupant and times of operation. All this data was then collected and analysed by specialists in Munich and New York who then estimated the technical effectiveness of RFE/RL broadcasts into the target zones.

Transmitting stations

Biblis

Located east of the German town of Biblis, 26 km north of Mannheim, the site's coordinates are 49.24N × 8.29E and it occupies 120 acres (48 Ha). An additional area of 216 acres (87 Ha) to the east is clear area to facilitate low angle radiation. The station has ten 100 kW SW transmitters, ten high gain curtain arrays and two rhombics. Seven of the curtains are diplexed to permit operation on two adjacent frequency bands. Two other curtains are triplexed permitting simultaneous operation in three adjacent frequency bands. The tenth curtain is triplexed to permit either 7 or 9 MHz with 11 MHz simultaneously. Two of the curtains are slewable in both vertical and azimuthal bearing.

All ten transmitters are fed programme material from master control in Munich over a ten-channel leased satellite facility, with ten dial-up telephone lines in the event of satellite failure. Additionally, eight programme lines (four in each direction) are installed between Biblis and Lampertheim to permit transfer and further back-stopping of programme material between these two stations.

Lampertheim

The Lampertheim SW transmitter station is located at 49.36N × 8.33E, 30 km northwest of Heidelburg in Germany and occupies 166 acres (67 Ha) of the State of Hessen and Forestry Department land. It has nine SW transmitters, each of 100 kW carrier power. These are connected through an antenna matrix to six 4 stack × 2 bay curtains, also with diplexers; all are orientated to target the former USSR. A seventh antenna is a dipole with a screen reflector to beam broadcasts to the former Democratic Republic of Germany. Programme material for Lampertheim was produced in the Munich studios and reached the transmitting station over ten satellite programme channels. For back-up ten special telephone lines were available in the event of failure.

Holzkirchen

Again in Germany, this transmitting station is located at 47.52N × 11.43E, east of the town and approximately 33 km southeast of Munich. Programmes produced in Munich are sent to the transmitter station over 10 digital audio circuits. As a back-up these programmes could be relayed over HF ISB transmitters to the Portuguese SW station. Some of the programmes could be transmitted at the same time from Holzkirchen to the target areas on either SW or MW. The programme relay to Portugal used seven ISB transmitters (10 kW) and carried a total of 14 programme channels. Holzkirchen has four 250 kW SW transmitters and four curtain arrays, and a 150 kW MW transmitter with a four-tower directional antenna to provide a Polish service.

Gloria

This situation is near the town of Gloria in Portugal, 65 km northeast of Lisbon, with coordinates 39.02N × 8.39W. Gloria was the largest RFE/RL transmitting station, equipped with nineteen 250 kW and two 50 kW SW transmitters. These fed through a cross-bar matrix to 6 stack high, highly directive curtains with forward gains ranging to 22.5 dB, equal to a power gain of nearly 200 and diplexed or triplexed so as to accept outputs from 2 or 3 transmitters each of 250 kW carrier output. The antenna field comprised 21 curtain arrays and 10 rhombic antennas in total.

Maxoqueira

This was the newest of the RFE/RL SW stations, completed in 1990–1991. The property is an approximate rectangle of 324 acres (141 Ha) about 600 m inland and near to the town of Benevente. Its geographical coordinates are 38.57N × 8.45W. The area is half wooded, mainly with cork trees. The transmitting site is an extension of the Maxoqueira receiving site taken out of service on 30th September 1989.

It was equipped with six SW transmitters, all of 500 kW carrier power rating, supplied by ABB. These transmitters were of the latest generation PSM design – the most modern and highest powered SW transmitters in the entire RFE/RL transmitter network. Unfortunately they were too late to play a role in the cold war because the station was only completed in 1991 after the abrupt, and totally unexpected, abandonment of Soviet

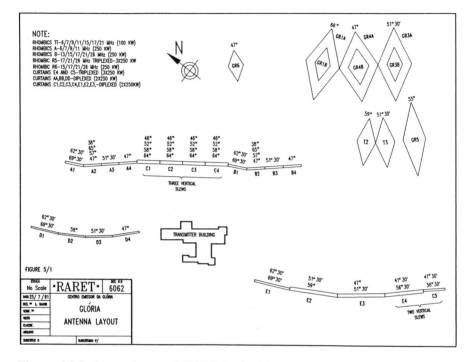

Figure A1.3 *Antenna layout at RFE/RL Station Gloria, Portugal*

communism. As explained elsewhere, it takes several years to plan and construct a high power SW facility in a foreign country, during which time much can happen in world politics.

Maxoqueira was equipped with six curtains suspended from seven towers and full flexibility of connection between transmitters and antennas was possible through a cross-bar switching matrix. Each antenna had a triplexer allowing it to accept three programmes on separate frequencies, with carrier powers of 250 kW, or two programmes simultaneously at 500 kW. Programme feeds to Maxoqueira were relayed from the Gloria transmitting station via a microwave link.

Pals

The Pals SW transmitting station is located in Spain. Located at 41.59N × 3.12E, the site at Playa de Pals lies 120 km north of Barcelona and occupies

82 acres (33 Ha) on a rectangular piece of land bordered on one side by the Mediterranean Sea.

The station broadcasts from six high power SW transmitters, each with a carrier power output of 250 kW. These transmitters feed through a switch system and transmission lines to nine curtain antennas arranged into four groups. Eight are designed to operate on one of two adjacent bands each, and four antennas are diplexed to allow simultaneous operation on adjacent bands, i.e. a 1000 kW carrier power into the one curtain.

All the antennas are of the curtain array type and of high performance with slewing facilities, mostly in both azimuth and vertical angle. Three are 4 bay wide × 8 stack, one is 2 bay × 4 stack, one is 4 bay × 4 stack and three are 6 bay × 4 stack, and most are capable of diplex operation, accepting two 250 kW inputs. The forward gains of these antennas range up to 23.26 dB.

The programme feeds to this station came from a 2.048 Mbit/s digital audio system via the EUTELSAT 1-F5 satellite; the signals are received on a 4.5 m satellite dish at the Playa de Pals site and then converted down into three channels of programme audio and then fed to all six SW transmitters.

This Appendix has described the network of RFE/RL at its peak, at the end of the cold war with the Soviet Union and the liberation of Eastern Europe in 1991. As mentioned in preceding chapters, some significant changes have taken place between 1991 and the present day. For example, the European Headquarters of RFE/RL has been relocated from Munich to Prague in the Czech Republic. There have also been other changes in the transmission infrastructure. The SW sites at Gloria and Maxoqueira in Portugal have been dismantled. Maxoqueira's 500 kW SW transmitters and SW antennas have been redeployed on Tinian Island in the Northern Marianas. Both of these sites are of considerable interest to those wishing to know more about the cold war and whilst we may not able to preserve such monuments, at least we can record them before they vanish forever.

Appendix 2
SW reference list for the major transmitter manufacturers

The following tables list SW transmitters of 50 kW power and above supplied by the major manufacturers Thomcast, Telefunken, GEC-Marconi and Continental Electronics, for the period 1980–1996. The data are reproduced with the permission of the manufacturers.

Country	Broadcaster or Network Operator	Station	Qty	Power (kW)	Type	On air
Thomcast						
Eritrea	Voice of the Broad Masses of Eritrea	Selea Daro	2	100	TRE 2315	1996
France	TDF – TéléDiffusion de France	Issoudun	4	500	TRE 2355	1996
Italy	Radio Vaticana	Santa Maria di Galeria	1	500	SK 55 C3-3P5	1996
Spain	RNE – Radio Nacional de Espana	Noblejas	2	250	TSW 2250	1996
Egypt, Arab Republic of	ERTU – Egyptian Radio & TV Union	Abu Zaabal	2	500	SK 55 C3-3P5	1995/1996
Nigeria, Fed Republic of	National Broadcasting Commission	Ikorodu	5	250	SK 53 C3-3P5	1995/1996
Egypt, Arab Republic of	ERTU – Egyptian Radio & TV Union	Abu Zaabal	2	500	TSW 2500	1995
France	TDF – TéléDiffusion de France	Issoudun	4	500	TRE 2355	1995
Kuwait	MOI – Ministry of Information	Kabd	2	500	SK 55 C3-3P5	1995
Kuwait	MOI – Ministry of Information	Kabd	1	500	SK 55 C3-3P5	1995
Libyan Arab Jamahiriya	LJB – Libyan Jamahiriya Broadcasting (via EGC)	Sabrata (Tripoli)	2	500	TRE 2355	1995
São Tomé e Principe	USIA/VOA – Voice of America	VOA-São Tomé Relay	4	100	SK 51 C3-3P6	1995
Australia	ABC – Australian Broadcasting Corp.	Darwin	2	250	TRE 2326	1994
France	TDF – TéléDiffusion de France	Issoudun	3	500	TRE 2355	1994
Guiana, French	RDF – TéléDiffusion de France	Montsinery	2	500	TRE 2355	1994
Libyan Arab Jamahiriya	LJB – Libyan Jamahiriya Broadcasting (via EGC)	Tripoli Sabrata	2	100	SK 51 C3-3P5	1994
Malaysia, Fed. of	RIM – Radio Television Malaysia	Kuching (Stapok/Sarawak)	1	100	SK 51 C3-3P5	1994
Canada	CBC – Canadian Broadcast Corporation	Sackville	3	250	SK 53 C3-3P5	1993/1995
Australia	ABC – Australian Broadcasting Corp.	Darwin	2	250	TRE 2326	1993/1994
France	TDF – TéléDiffusion de France	Issoudun	1	500	TRE 2355	1993
Iran	IRIB – Islamic Republic of Iran Broadcasting	Mashhad	2	250	SK 53 C3-2P	1993
Korea, Republic of	KBS – Korean Broadcasting System	Kimje III	1	250	SK 53 C3-3P5	1993
Sweden	Sevensk Runradio AB	Hörby	3	500	SK 55 C3-3P5	1993
Kuwait	MOI – Ministry of Information	Kabd	2	500	SK 55 C3-2P	1992
Philippines, Republic of	Radio Veritas Asia	Palauig III	1	250	SK 53 C3-2P	1992
Rwanda, Republic of	DW – Deutsche Welle	DW-Kigali	2	250	SK 53 C3-2P	1992
Saudi Arabia	MOI – Ministry of Information	Riyadh-SW	2	500	TRE 2355	1992

Thomcast

Country	Broadcaster or Network Operator	Station	Qty	Power (kW)	Type	On air
Switzerland	PTT – Direction Générale de L'Entreprise des PTT	Beromünster	1	250	SK 53 C3-2P	1992
Singapore	SBC – Singapore Broadcasting Corporation	Kranji	1	100	SK 51 C3-2P	1991/1992
Singapore	SBC – Singapore Broadcasting Corporation	Kranji	6	250	SK 53 C3-2P	1991/1992
Turkey	TRT – Türkiye Radyo-Televizyon Kurumu	Emirler	5	500	SK 55 C3-2P	1991/1992
Rwanda, Republic of	Orinfor	Kininya	1	100	SK 51 F3-2P	1991
Rwanda, Republic of	DW – Deutsche Welle	DW-Kigali	2	250	SK 53 C3-2P	1991
Saudi Arabia	MOI – Ministry of Information	100 kW SW Station	3	100	TRE 2315	1991
South Africa, Republic of	SABC – South African Broadcasting Corporation	Langefontain	8	100	SK 51 C3-2P	1991
South Africa, Republic of	SABC – South African Broadcasting Corporation	Meyerton	4	100	SK 51 C3-2P	1991
Zaire	La Voix du Zaire	Masina (Kinshasa)	1	100	SK 51 F3-2P	1991
Kuwait	MOI – Ministry of Information	Kabd	6	500	SK 55 C3-2P	1990/1991
Portugal	RFE/RL – Radio Free Europe/Radio Liberty	Maxoqueira	6	500	SK 55 C3-2P	1990/1991
Egypt, Arab Republic of	ERTU – Egyptian Radio & TV Union	Abu Zaabal	1	100	TRE 2315	1990
Germany	RFE/RL – Radio Free Europe/Radio Liberty	Biblis	2	100	TRE 2315	1990
Germany	RFE/RL – Radio Free Europe/Radio Liberty	Lampertheim	1	100	TRE 2315	1990
India	AIR – All India Radio	Panaji	2	250	SK 53 C3-2P	1990
Iran	IRIB – Islamic Republic of Iran Broadcasting	Kamalabad	2	100	SK 51 C3	1990
Iran	IRIB – Islamic Republic of Iran Broadcasting	Kamalabad	2	250	SK 53 C3-2P	1990
Korea, Republic of	KBS – Korean Broadcasting System	Hwa Sung	3	100	SK 51 F3	1990
Libyan Arab Jamahiriya	LJB – Libyan Jamahiriya Broadcasting (via ECG)	Sabrata (Tripoli)	1	500	SK 55 C3	1990
Switzerland	PTT – Direction Générale de L'Entreprise des PTT	Sottens	1	500	SK 55 C3-2P	1990
USA	USIA/VOA – Voice of America	Bethany (OH)	3	250	SK 53 C3	1990
Saudi Arabia	MOI – Ministry of Information	Riyadh-SW	4	500	TRE 2355	1989/1991
India	All India Radio (via BEL)	Gorakhpur	1	50	SK 45 F3	1989/1990
India	All India Radio (via BEL)	Itanagar	1	50	SK 45 F3	1989/1990
India	All India Radio (via BEL)	Srinagar	1	50	SK 45 F3	1989/1990

Country	Organization	Location		Power	Type	Year
India	All India Radio (via BEL)	Timpu	1	50	SK 45 F3	1989/1990
India	AIR – All India Radio	Bangalore	4	500	SK 55 C3-2P	1989/1990
Ethiopia	Radio Ethiopia	Jewe	1	100	SK 51 F3	1989
Gabon	Africa N° 1	Moyabi 2	1	500	TRE 2355	1989
Italy	RAI – Radio Televisione Italiana	Prato Smeraldo	4	100	SK 51 C3-2P	1989
New Zealand	RNZ – Radio New Zealand	Rangitaiki	1	100	TRE 2315	1989
Tanzania	Radio Tanzania	Mabibo (Dar es Salaam)	2	100	SK 51 F3	1989
USA	WSHB – Monitor Radio International	WSHB-Cypress Creek (AL)	2	500	SK 55 C3-2P	1989
Guiana, French	TDF – TéléDiffusion de France	Montsinery	1	500	TRE 2355	1988
India	AIR – All India Radio	Khampur (Delhi)	2	250	SK 53 F3	1988
India	AIR – All India Radio	Bangalore	2	500	SK 55 C3	1988
Jordan	JRTV – Jordan Radio & Television Corp.	Qasr Kherane	3	500	SK 55 C3-2P	1988
Kuwait	MOI – Ministry of Information	Kabd Sulabiyah	2	500	SK 55 C3-2P	1988
Netherlands Antilles	Radio Nederland Wereldomroep	Bonaire	1	250	SK 53 C3-2P	1988
Philippines, Republic of	Radio Veritas Asia	Palauig II	1	250	SK 53 C3	1988
Yemen, Republic of	Yemen Radio & TV Corporation	Sanaa	1	300	TRE 2335	1988
Hungary	Magyar Radio	Szekesfehervar	2	100	SK 51 C3-2P	1987/1988
Gabon	Radiodiffusion Télévision Gabonaise	Melen (Libreville)	1	100	TRE 2311P	1987
Korea, Republic of	KBS – Korean Broadcasting System	Seoul	3	100	SK 51 F3	1987
USA	Christian Science Publishing Society Boston	WCSN-Greenbush (MA)	1	500	SK 55 C3-2P	1987
Iran	IRIB – Islamic Republic of Iran Broadcasting	Kamalabad	6	500	SK 55 C3-2P	1986/1988
Iran	IRIB – Islamic Republic of Iran Broadcasting	Mashhad	4	500	SK 55 C3-2P	1986/1988
Côte d'Ivoire	Radiodiffusion Télévision Ivoirienne	Akuedu (Abidjan)	1	500	TRE 2355	1986
Finland	Oy Yleisradio AB	Pori II	1	100	SK 51 C3-2P	1986
India	AIR – All India Radio	Guwahati	1	50	SK 45 F3	1986
India	AIR – All India Radio	Khampur (Delhi)	2	50	SK 45 F3	1986
India	AIR – All India Radio	Kingsway (Delhi)	2	50	SK 45 F3	1986
India	AIR – All India Radio	Shillong	1	50	SK 45 F3	1986
Iran	IRIB – Islamic Republic of Iran Broadcasting	Kamalabad	2	500	SK 55 C3-2P	1986
Italy	Radio Vaticana	Santa Maria di Galeria	2	500	SK 55 C3-2P	1986
Korea, Republic of	KBS – Korean Broadcasting System	Kimje II	1	250	SK 53 C3-2P	1986
Philippines, Republic of	Radio Veritas Asia	Palauig I	1	250	SK 53 C3-2P	1986

Country	Broadcaster or Network Operator	Station	Qty	Power (kW)	Type	On air
Thomcast						
Switzerland	PTT – Direction Générale de L'Entreprise des PTT	Schwarzenburg	1	100	SK 51 C3	1986
USA	USIA/VOA – Voice of America	Delano (CA)	4	250	SK 53 C3	1986
USA	USIA/VOA – Voice of America	Greenville (NC)	1	500	SK 55 C3	1986
USA-Guam	AWR – Advendist World Radio-Asia	Guam	2	100	TRE 2311 P	1986
Yugoslavia, Fed Rep of	JRT – Udruzenje Jugoslovenskih Radiotelevizija d.O.O.	Bijeljina	4	500	SK 55 C3	1986
Finland	Oy Yleisradio Ab Helsinki	Pori II	3	500	SK 55 C3-2P	1985/1986
Iran	IRIB – Islamic Republic of Iran Broadcasting	Kamalabad	2	500	SK 55 C3	1985
United Arab Emirates	Ministry of Information & Culture	Dabiya (Abu Dhabi)	4	500	SK 55 C3	1985
Australia	ABC – Australian Broadcasting Corp.	Carnarvon	1	300	TRE 2330	1984
Belgium	BRTN – Belgische Radio en Televisie	Wavre	1	100	SK 51 C3	1984
Guiana, French	TDF – TéléDiffusion de France	Montsinery	3	500	TRE 2352	1984
India	AIR – All India Radio	Bombay	1	100	SK 51 F3	1984
India	AIR – All India Radio	Madras	1	100	SK 51 F3	1984
Iran	IRIB – Islamic Republic of Iran Broadcasting	Kamalabad-Karaj	4	500	SK 55 C3	1984
Kenya	KBC – Kenya Broadcasting Corporation	Koma Rock (Nairobi)	2	250	TRE 2320	1984
Iraq	MOI – Ministry of Information	Balad	16	500	TRE 2352	1983/1984
Bangladesh	Radio Bangladesh	Kabirpur (Dhaka)	2	250	TRE 2320	1983
Cameroon	C.R.T.V. – Cameroon Radio Television	Douala	1	100	TRE 2310	1983
Chad	Radiodiffusion National Ichadienne	Gredia (N'Djamina)	1	100	TRE 2310	1983

Country	Organization	Location	No.	Power	Type	Year
Hungary	Magyar Radio	Diosd	2	100	SK 51 F3	1983
India	AIR – All India Radio	Kingsway (Delhi)	2	100	SK 51 F3	1983
India	AIR – All India Radio	Aligarh	2	250	SK 53 F3	1983
Indonesia	RRI – Radio Republik Indonesia	Cimanggis (Jakarta)	1	250	TRE 2320	1983
Indonesia	RRI – Radio Republik Indonesia	Medan	1	250	TRE 2320	1983
Iraq	MOI – Ministry of Information	Abu Ghraib	4	250	SK 53 C3	1983
Libyan Arab Jamahiriya	LJB – Libyan Jamahiriya Broadcasting	Sabrata (Tripoli)	2	500	SK 55 C3	1983
Morocco	MEDI 1 – Radio Mediterranée Internationale	Nador	2	250	TRE 2320	1983
Turkey	TRT – Türkiye Radyo-Televizyon Kurumu	Cakirlar-Ankara	2	500	SK 55 C3	1983
Denmark	Telecom A/S	Herstedvester	1	100	SK 51 F3	1983
France	RMC – Radio Monte Carlo	Fontbonne	1	500	TRE 2351	1982
Italy	RAI – Radio Televisione Italiana	Prato Smeraldo	2	100	SK 51 C3	1982
Nigeria, Fed Republic of	National Broadcasting Commission	Ikorodu	2	500	SK 55 F3	1982
Philippines, Republic of	USIA/VOA – Voice of America	Tinang	2	250	SK 53 C3	1982
Syrian Arab Republic	Organisme de la Radio-Télévision Arabe Syrienne	Adra (Damascus)	4	500	TRE 2352	1982
Kuwait	MOI – Ministry of Information	Kabd Sulabiyah	2	500	SK 55 F3	1981
Malaysia, Federation of	RTM – Radio Television Malaysia	Kajang (Penang)	2	500	TRE 2351	1981
Mauritania, Islamic Republic of	Radio Mauritanie	Nouakchott	1	100	SK 51 F3	1981
Nigeria, Fed Republic of	National Broadcasting Commission	Katabu	2	100	TRE 23211	1981
Iraq	MOI – Ministery of Information	Babylone	2	500	TRE 2352	1980/1981

Country	Broadcaster or Network Operator	Station	Qty	Power (kW)	Type	On air
Telefunken Sendertechnik						
Germany	Telekom	Nauen	4	500	S4105	1996/1997
Norway	Telenor	Sveio	1	500	S4105	1996
Iran	IRIB	Sirjan	10	500	S4005	1990/1995
Portugal	RDP CEDC	S. Gabriel	1	300	S4005	1989
PR Congo			2	100	S4001	1989
Germany	DBP, DW	Wertachtal	6	500	S4005	1987/1989
Austria	ORF	Moosbrunn	1	500	S4005	1987
Japan	KDD	Yamata	2	300	S4005	1987
Israel	Bezeq	Hillel	1	500	S4005	1986
Norway	NTA	Sveio	1	500	S4005	1986
USA	VOA	Greenville, NC	1	500	S4005	1986
Sri Lanka	DW	Trincomalee	2	300	S4003	1984/1985
Germany	DBP, DW	Julich	10	100	S4001	1984/1988
Netherlands	PTT	Flevoland II	1	100	S4001	1984/1985
Vatican City	RV	S.M. di Galeria	1	500	S4005	1984
Netherlands	PTT	Flevoland II	4	500	S4005	1983/1984
Germany	RIAS	Berlin	1	100	S4001	1983
UK	BBC-WS	Rampisham	4	500	S4005	1982/1985
Norway	NTA	Kvitsøy	2	500	S4005	1982/1984
Austria	ORF	Moosbrunn	1	300	S4005	1982/1983
Germany	DBP, DW	Wertachtal	1	500	SV2500	1982/3
Germany	BR	Ismaning	1	250	S4005	1982

Country	Broadcaster or Network Operator	Station	Qty	Power (kW)	Type	On air
GEC-Marconi						
Egypt	ERTU		1	500	B6132	1996/1997
Indonesia	RRI		3	250	B6131	1996
Indonesia	RRI		9	250	B6131	1993
USA	VOA		4	500	B6128	1993
USA	VOA		17	500	B6128	1992
Cyprus	BEMRS		2	250	B6131	1991
UK	BBC-WS		2	500	B6128	1990
Seychelles	BBC-WS		2	250	B6131	1987
UK	BBC-WS		3	300	B6126	1986
Spain	RFE		1	250	B6131	1986
Hong Kong	BBC-WS		2	250	B6131	1986
UK	FCO		7	300	B6126	1985
US	VOA		1	500	B6127	1985
UAE	DRC TV		1	500	B6127	1985
UK	BBC		4	500	B6127	1984
Cyprus	DWS		4	300	B6124	1981
Far East			3	300	B6124	1980
UK	BBC		4	300	B6124	1980

Country	Broadcaster or Network Operator	Station	Qty	Power (kW)	Type	On air
Continental Electronics Corporation						
India	All India Radio	Delhi	1	250	419H	1996
China	Ministry Radio, Film & TV	Beijing	8	500	420C	1995
Qatar	Ministry of Information & Culture	Al-Khassah	1	500	420C	1995
India	All India Radio	Delhi	1	250	419H	1995
China	Ministry Radio, Film & TV	Beijing	12	100	418F	1995
Guam	Adventist Broadcast Service		1	100	418F	1995
USA	WWCR	Nashville, TN	1	100	418F	1995
China	Ministry Radio, Film & TV	Beijing	2	500	420C	1994
China	Ministry Radio, Film & TV	Beijing	2	100	418F	1994
Zimbabwe	ZBC	Gweru	2	100	418E	1994
Zambia	Christian Vision	Lusaka	1	100	418E	1994
Guam	Adventist Broadcast Service		1	100	418E	1994
India	All India Radio	Delhi	1	50	417E	1994
Laos	Telecom Australia		1	50	417E	1994
USA	WWCR	Nashville, TN	3	100	418E	1993
Papua, New Guinea	NBC	Port Moresby	2	100	418E	1993
USA	Caribbean Beacon	Dallas, TX	1	100	418E	1993
USA	Eternal Word	Birmingham, AL	4	500	420C	1992
Taiwan	Central Broadcasting Station	Taipei	3	100	418E	1992
West Indies	Caribbean Beacon	Anguilla	1	100	418E	1992
Botswana	VOA	Selebe-Phikwe	4	100	418E	1991
China	Ministry Posts & Telecom	Xinjiang	4	100	418E	1991
China	Ministry Radio, Film & TV	Yunnan	2	100	418E	1991

Country	Organization	City	Qty	Model	Year
Philippines	FEBC	Manila	1	418E	1991
Israel	Ministry of Communication	Yaune	1	420B-1	1989
Costa Rica	Radio Nacional Espana	San Jose	3	418D-2	1989
Saipan	Herald (Christian Science)		1	419F-2	1989
Portugal	RFE	Gloria	8	419F-2	1987/1988
Korea	Korean Telecom (KBS)	Seoul	2	418D-2	1987
USA	VOA	Greenville, NC	1	420B	1985
Saipan	FEBC		2	418D-2	1985
Philippines	FEBC	Manila	1	418D-2	1985
Lesotho	Radio Lesotho	Maseru	1	418D-2	1985
Germany	RFE	Lampertheim	3	418D-2	1984
Germany	RFE	Biblis	1	418D-2	1984
Australia	Telecom Australia	Katherine	1	418D-2	1984
Australia	Telecom Australia	Alice Springs	1	418D-2	1984
Australia	Telecom Australia	Tennant Creek	1	418D-2	1984
Taiwan	Broadcasting Corp. of China	Taiwan	2	418D-2	1983
Israel	Ministry of Communication	Yavne	1	418D-2	1983
Saipan	Herald (Christian Science)		1	418D-2	1983
Saipan	KFBS/FEBC		1	418D-2	1983
Tunisia	Middle East Wire & Wireless		1	418D-2	1983
Oman	Radio Oman	Salalah	1	418D-2	1982
Saipan	FEBC		1	418D-2	1982
India	All India Radio	Delhi	1	417E	1981
Egypt	ERTU	Abu Zaabal	1	419F	1980
Egypt	Ministry of Defence	Cairo	1	417D	1980

Index